ISOTOPES AND RADIATION IN BIOLOGY

ISOTOPES AND RADIATION IN BIOLOGY

C.C. THORNBURN, B.Sc., Ph.D.
Lecturer
The Department of Biological Sciences
The University of Aston in Birmingham
Great Britain

Halsted Press Division
JOHN WILEY & SONS, INC.
NEW YORK

English edition first published in 1972
by Butterworth & Co (Publishers) Ltd
London; 88 Kingsway, WC2B 6AB

Published in the United States and Canada
by Halsted Press Division, John Wiley & Sons Inc.
New York

Library of Congress Cataloging in Publication Data
Thornburn, C C
Isotopes and radiation in biology.
Includes bibliographies.
1. Radiobiology. I. Title.
QH652.T45 574.1'915 72-2293
ISBN 0-470-86184-3

Printed in Hungary

PREFACE

One of the first tasks I was asked to undertake when I arrived at The University of Aston in Birmingham was the development, as part of the degree in Biological Sciences, of an undergraduate course on the subject of radiation and isotopes in biology, consisting of some ten lectures and associated practical work. I found that the textbooks covering the field could be divided into two groups. On the one hand were the inexpensive paperbacks, aimed at the legendary man-in-the-street, very readable and giving an easily understood general introduction, but rather too simplified for a university course. On the other hand were the specialised texts intended for researchers which, though excellent, were very detailed and rather too high-powered for undergraduates and, as a consequence of this, were large and expensive, particularly for only a minor part of the course. To make matters worse, no one book covered all aspects of the subject, and three costly volumes would have been needed: one on the use of isotopes, another on the effects of radiation on living things, and a third on radioactivity in the living environment. Thus there seemed to be a gap in the literature, for I did not come across a single book that was intermediate between these two groups and which surveyed the whole field. The obvious solution was suggested to me, indeed almost suggested itself—'Well then, why don't you write it yourself?' So I did—and that is the reason for the existence of this book. I would have liked a short simple title, but have had to settle for a somewhat pedestrian one, for there is no single name covering the subject of ionising radiation and biology as a whole, radiation both as a tool and a hazard.

To make it easy for the reader to find his way about, each chapter in the book is divided into numbered sections, subdivided as necessary, with section headings and key words and phrases emphasised. The sections should not be regarded as mutually exclusive watertight compartments, for it is often the case that knowledge of one topic is relevant to full understanding of another; I have treated such topics fully in what I felt was the logical place, with cross-references as appropriate from other sections in which the topic is mentioned. The literature references at the ends of the chapters include other textbooks, and a selection of recent research papers to indicate current developments and to serve as starting points for literature searches if more information is desired. I have assumed that readers of the book will have a basic knowledge of chemistry and biology, and therefore I have not explained what is meant by concepts such as the Periodic Table of the elements and mitotic cell division.

I have tried not to sacrifice accuracy and completeness to brevity, and I make no apology for having been brief and concise; this is for the sake of my student readers, both to cut down the size and cost of the book, and also because no one relishes wading through masses of verbiage to try to discover the point the author is making. I trust that my book will help them to find their way through what is in parts a complex and confusing subject.

Here then is what I hope will be 'radiobiology made easy', or at least 'radiobiology without tears'.

ACKNOWLEDGEMENTS

I am grateful to the United Kingdom Atomic Energy Authority for permission to reproduce Figure 6.1, to Professor P. Alexander and Pergamon Press for Figure 7.1 and to Dr. B. Åberg and Pergamon Press for Figure 8.1. My thanks are also due to Mrs. O. Deeley, who typed most of the manuscript, and to the many other people who have helped directly or indirectly in the preparation of this book: since it is based largely on lecture notes, it would be difficult or impossible to acknowledge all sources individually.

Birmingham C.C.T.

CONTENTS

Chapter 1

INTRODUCTION

1.1 ARRANGEMENT OF THE SUBJECT

As 'Gallia in tres partes divisa est', to quote yet again that well-worn statement of the great Julius Caesar, so is this book on the study of ionising radiation in biology. The first section gives a brief review of the history of the discovery of radiation, which may be called radio-history (Chapter 1), and an outline of the physical aspects of radiation, or radio-physics (Chapter 2).

The two major sections following this introduction cover, firstly, the use of radiation as a tool in biology, subdivided into the detection, the usage, and the uses of radiation (Chapters 3 to 5). The detection and usage of radiation together form the topic of the isotopic tracer technique, sometimes known in the New World as radiotracer methodology. Its uses include radiography, radiotherapy and radiosterilisation. The subject matter of this first section is relatively straightforward, being based largely on the physical sciences, in contrast to the second main section, which is concerned with what happens when ionising radiations interact with living systems, cells and organisms. This latter is a complex field, by no means fully understood, and therefore contains much uncertainty: probability and statistical significance are prominent, rather than definite statements. Here again there are three subdivisions: the control, the effects, and the ecology of radiation as a hazard to life (Chapters 6 to 8), called respectively health physics, radiation biology or radiobiology, and radio-ecology. The book concludes with an indication of some possibilities for class experiments to illustrate the theory outlined in the text. The production of radio-isotopes is not dealt with in detail, being of direct interest more to the physicist than to the biologist.

1.2 RADIO-HISTORY

The story of radioactivity belongs almost entirely to the twentieth century, and with the modern developments of nuclear power it may very easily be thought to be a very new thing. But although such developments are entirely of this day and age, radioactivity itself is very old, having been in existence since the Earth was created, though how it was initially produced will not be discussed here.

Radioactivity's discovery has one happy and somewhat unusual feature, that it can be followed in detail. Its discoverer, the French physicist Henri Becquerel, was a member of the Académie des Sciences in Paris, and was thereby in the fortunate position of being able to publish his findings almost as they happened, like a series of progress reports, at the (then) weekly 'Séances' of the Académie, whose proceedings were in print within ten days: most scientists then (as now) had to wait months for publication of their papers. By reading through 'Compte Rendu' of the Académie for the first few months of 1896, one can become almost an eye-witness of the discovery of radioactivity, and retrace the steps that led to the firm establishment of the existence of this new phenomenon.

In 1895, Wilhelm Conrad Röntgen discovered the rays that in German-speaking countries now bear his name and which are more universally known as x-rays. They aroused tremendous scientific and public interest, since they could penetrate where the eye could not see and reveal what lay beneath the surface, as for example the bones in a living hand. They were produced when cathode rays struck the wall of a discharge tube: at the same spot the glass fluoresced brightly, and this led Becquerel to wonder whether x-rays might also be produced when a fluorescent material was exposed to light. Accordingly he placed some fluorescent crystals—of potassium uranyl sulphate—on top of a photographic plate wrapped in black paper, and exposed them to the sun for several hours. When he developed the plate, it showed the hoped-for result: 'On doit donc conclure de ces expériences que la substance phosphorescente en question émet des radiations qui traversent le papier opaque à la lumière et réduisent les sels d'argent'. ('One can conclude from these experiments that the phosphorescent substance in question emits radiations which penetrate paper opaque to light and reduce the salts of silver.')

And, but for the weather, that might have been that. However—'Parmi les expériences que précèdent, quelques-unes avaient été préparées le mercredi 26 et le jeudi 27 février, et comme les jours-là le soleil ne s'est montré que d'une manière intermittente, j'avais conservé les expériences toutes préparées et retiré les châssis à l'obscurité dans le tiroir d'un meuble, en laissant en place les lamelles du sel d'uranium. Le soleil ne s'étant pas montré de nouveau les jours suivants, j'ai développé les plaques photographiques le 1er mars, en m'attendant à trouver des images très faibles. Les silhouettes apparurent, au contraire, avec une grande intensité.' ('Among the preceding experiments, some were prepared on Wednesday, 26 and Thursday, 27 February and, as on those days the sun appeared only intermittently, I held back the experiments that had been prepared and returned the plate-holders to darkness in a drawer, leaving the crystals of the uranium salt in place. As the sun still did not appear during the following days, I developed the photographic plates on the first of March, expecting to find very weak images. On the contrary, the silhouettes appeared with great intensity.') What made Becquerel develop those plates we do not know—most people would probably have thrown them away—but there it was, an apparently abortive experiment had produced a most unexpected and fruitful result and, without knowing it, Becquerel had made the first autoradiograph. As he looked more closely into the nature of his 'radiations invisibles', he found that they seemed to have less and less connection with light and fluorescence, and there seemed to be an inexhaustible supply of energy to produce them. A crystal of uranium nitrate, dissolved and remade in the dark, should have had a structure free from any energy due to light, yet it radiated as strongly as ever; so did the solution, which neither phosphoresced nor fluoresced. By 18 May 1896, the discovery was complete; that here indeed was a new thing, for which the only thing necessary was the presence of uranium. For want of a better term, and as he had not altogether abandoned hope that the radiation was in some way brought about by means of light, Becquerel called his radiation 'invisible phosphorescence'.

The discovery of radioactivity is typical of other discoveries in that it was made by chance; Becquerel's original experiment would have been most unlikely to have succeeded in the way he intended in view of the relative energy contents of light and x-rays. As the story unfolded, it proved to be full of surprises and (as Romer has put it) an eye for the unexpected was worth quite as much as any foresight in planning. The history of radioactivity also shows how a discovery has to wait for development of the means by which it can be discovered, though it is conceivable that radioactivity could have been found earlier by means of the electroscope or the electrometer, as it shares with x-rays the property of being able to discharge electrified bodies. However, before it could be discovered in the way it was, there had to be photographic emulsions that could record penetrating radiation, and the discharge tube to produce such radiations and establish a connection between them and fluorescence. Finally, the right fluorescent material had to be chosen for the crucial experiment, one that was also radioactive: an organic substance would have been no use, the experiment would have failed, and the work might well have been abandoned with the conclusion that: 'Though fluorescence is associated with the production of x-rays, they are not produced by fluorescent materials'. But, given the juxtaposition of uranium and a photographic emulsion, it seems inevitable that radioactivity would have been discovered sooner or later, and it was perhaps sheer good luck that Becquerel found it as soon as he did and before anyone else tried the same experiment.

Working with Becquerel at this time was a remarkable young lady from Poland, Marie Sklodowska, who did several things that were unheard of for a woman in those days, when the sexes were even less equal than they are now. She taught herself science, and in 1891, at the tender age of 24, she journeyed to Paris to study mathematics and physics, where four years later she married her physics professor. She then began to work for her doctorate, continuing Becquerel's work on the new rays by examining all known elements for their presence. It proved at first a rather disappointing study for, out of the eighty-odd known elements, the only one (other than uranium) which was active enough to warrant further study was thorium.

Something else, however, was rather more promising. All compounds of uranium and thorium would be active, of course, and it was to be expected that the activity would be less than that of the pure element according to its proportion in the compound. Not so the uraniferous minerals pitchblende and chalcolite—here was her breakthrough, the quantification of their activity revealing that pitchblende was $2\frac{1}{2}$ times as active as it should have been on the basis of its uranium content. 'Deux minéraux d'uranium, la pechblende et la chalcolite, sont beaucoup plus actifs que l'uranium lui-même. Ce fait est très remarquable et porte à croire que ces mineraux peuvent contenir un élément beaucoup plus actif que l'uranium. J'ai reproduit la chalcolite ... avec les produits purs; cette chalcolite artificiel n'est pas plus active qu'un autre sel d'uranium.' ('Two minerals of uranium, pitchblende and chalcolite, are much more active than uranium itself. This fact is most remarkable and leads one to believe that these minerals may contain an element which is much more active than uranium. I have made up chalcolite ... from pure materials, and this artificial chalcolite is no more active than any other salt of uranium.') Pierre Curie now joined his wife, and together they set to work to find out more about this intriguing element. Their report 'Sur une substance nouvelle radio-active contenue dans la pechblende', appears to contain the first use of the new word 'radio-active', and gives an account of the first radiochemical separation, i.e. a chemical separation whose progress is followed by means of radioactivity. 'Nos recherches chimiques ont été constamment guidées par la contrôle de l'activité radiante des produits séparés à chaque opération ... qui rende compte de la richesse du produit en substance active.' ('Our chemical research has been constantly guided by the radiating activity of the products separated at each operation ... which gives a measure of the richness of the product in an active substance.')

The new element came out analytically in the copper group (IIA) and a new name was modestly proposed for it. 'Le corps actif reste avec le bismuth ... et son activité est environ 400 fois plus grande que celle de l'uranium.... Nous croyons donc que la substance que nous avons retirée de la pechblende contient un métal non encore signalé, voisin du bismuth par ses propriétés analytiques. Si l'existence de ce nouveau métal se confirme, nous proposons de l'appeler *polonium*, du nom du pays d'origine de l'un de nous.' ('The active body came down with bismuth ... and its activity is about 400 times greater than that of uranium.... We believe, therefore, that the substance which we have extracted from pitchblende contains a new metal, hitherto unknown, very like bismuth in its analytical properties. If its existence is confirmed, we propose to call it polonium, after the native country of one of us.') They noted that: 'Cette découverte sera uniquement due au nouveau procédé d'investigation que nous fournissent les rayons de Becquerel.' ('This discovery is due entirely to the new procedure which the Becquerel rays give us.')

Six months later, another element appeared, this time in Group IV. 'Nous avons rencontré une deuxième substance fortement radio-active et entièrement différente de la première par ses propriétés chimiques.... La nouvelle substance que nous venons de trouver a toutes les apparences chimiques de baryum: néanmoins les diverses raisons nous portent à croire que la nouvelle substance radio-active renferme un élément nouveau, auquel nous proposons de donner le nom *radium*.' ('We have found a second strongly radioactive substance, entirely different from the first in its chemical properties.... The new substance which we have just found has all the chemical signs of barium: nevertheless, for a number of reasons we believe that the new radioactive substance contains a new element, to which we propose to give the name radium.') It was different not only chemically—'La nouvelle substance radioactive renferme certainement une très forte proportion de baryum: malgré cela, la radio-activité est considérable. La radioactivité du radium doit donc être énorme.' ('The new radioactive substance certainly contains a very large proportion of barium: in spite of this, its radioactivity is considerable. The activity of radium must therefore be enormous.') A third new element was found in 1900 by Debierne and Giesel, who were assisting the Curies; chemically similar to iron and uranium in Group IIIA, they called it *actinium*.

The Curies' work was eased through the gift of a large quantity of pitchblende residue from which the uranium had been extracted. ('Nous avons obtenu de 100 kg d'un résidu de traitement de pechblende de Joachimsthal (Autriche) ne contenant plus d'urane, mais contenant de polonium et du radium, que facilitera beaucoup nos recherches.') Even so, pitchblende contains only about 100 μg of polonium and 100 mg of radium per ton, and one pauses to reflect on the devotion of these pioneers when it is remembered that, in spite of the powerful tool of radioactivity for quality control, their separation methods were based on chemical group separation and fractional crystallisation, and they had none of the modern analytical aids that we take for granted. In recognition of their work, the two Curies and Henri Becquerel were jointly awarded the Nobel Prize for physics in 1903. Following Pierre's tragic death in a Paris street accident in 1906, Marie succeeded him as Professor of Physics, and went on to gain a second Nobel Prize, for chemistry, in 1911—the only person ever to receive two until Pauling gained the Peace Prize in 1963.

Interest in 'Becquerel's rays' had waned since 1896 in favour of the more spectacular Röntgen rays, but it was now reawakened by the Curie's discoveries, and several now-famous people promptly began to work on radioactivity themselves. Owens and Rutherford found that the activity of thorium, unlike uranium, was variable; the variability proved to be due to air currents, and they deduced that 'thorium continuously emits radioactive particles of some kind', which they called the *emanation*. The emanation seemed able to produce radioactivity on other things, and the amount of this excited radiation was completely independent of the material on which it was produced. There was no detectable increase in weight or alteration in appearance of the activated material, though the activity could be removed mechanically or chemically. The activities of the emanation and the excited radiation both fell away in geometrical progression with time, and this gave the first hint that radioactive elements were each characterised by a decay time factor.

The emanation we now know as the last inert gas; it was at first called niton, or named according to its origin, i.e. thorium-emanation or thoron, actinium-emanation or actinon, and radium-emanation or *radon*. All the emanations were found to be chemically the same, and the last name was subsequently adopted for all, to fit in with the other names in the inert gas group and to be a reminder that it is, in fact, radioactive.

Rutherford found in 1898 that uranium emitted rays of two distinct kinds. One, which he called *alpha*, had enormous ionising power, but would not penetrate through a sheet of paper; while the other kind, *beta*, ionised much less but travelled further. Villard discovered a third kind in 1900, *gamma*, which are even less ionising and far more penetrating. Later experiments by Rutherford (1911) on the scattering of alpha particles by thin metal foil showed that by far the greatest proportion passed through but, unexpectedly, a small number were deflected, by collision with the atomic nuclei. This result was very important in developing the theory that most of the atom is empty space, with the mass concentrated in a very small volume in the nucleus.

In 1900 Crookes made a puzzling discovery. Precipitating uranium from solution with ammonium carbonate and redissolving it in excess reagent left a small residue which contained all the original beta activity, the alpha activity remaining with the uranium. Thus uranium and its beta activity could be chemically separated, and Crookes proposed the existence of a 'uranium-X' to account for this. Rutherford and Soddy confirmed that the same was true for thorium, but more puzzling here was the fact that the mysterious 'X' gradually lost its activity, while the original thorium regenerated it, the sum of the two activities always being constant. They concluded that: 'The major part of the radioactivity of thorium is not due to the thorium at all, but to the presence of a non-thorium substance in minute amount which is being continuously produced.' There was nothing to produce 'thorium-X' except thorium, yet it was chemically different, and the only possible conclusion was that the one was produced by transmutation of the other, as Becquerel himself had concluded earlier about uranium. 'The position is thus reached that radioactivity is at once an atomic phenomenon and the accompaniment of a chemical change in which new kinds of matter are produced. The two considerations force us to the conclusion that radioactivity is a manifestation of subatomic chemical change ... the idea of the chemical atom in certain cases spontaneously breaking up with evolution of energy is not of itself contrary to anything that is known of the properties of atoms.... The changes brought to knowledge by radioactivity are of a different order of magnitude from any that have before been dealt with in chemistry.'

The Curies criticised the idea of transmutation, on the ground that radioactivity was a property of a chemical element and therefore permanent. They considered that the seemingly inexhaustible supply of energy of the radioactive elements was due to their being able to concentrate, in some unexplained way, an invisible radiation from their surroundings, which they could then re-emit or transfer to other substances. However, in spite of such formidable opposition Rutherford and Soddy thoroughly and carefully set about proving their theory. A slight, though profoundly significant, modification to it was the supposition that the radiation energy was released at the same instant as the atom transmuted, and it improved the transmutation theory so much that by 1904 it was fully accepted as the basis of radioactivity, for it answered so many questions so neatly. All the main basic features of radioactivity were now clear, and though much important work followed, it was largely a matter of unravelling all the details.

The main revelation was that uranium and thorium decay through long *decay chains* or series of some 10 or 14 steps, emitting alpha or beta particles and often gamma also, until they finally reach lead and stability. Uranium and thorium decay very slowly and gradually build up a chain in which, at equilibrium, each succeeding member is being formed as fast as it decays: the relative quantity of each member is proportional to its rate of decay, and is therefore constant, and from each atom of the *parent* (great-great... grandparent?), one atom of the final stable *daughter* is eventually formed. There are three *natural decay series*, sometimes called natural radioactive families. The atomic weights of the members of any one series differ by four all the way down, so that they can expressed as multiples of 4 plus a remainder. Thus the uranium–radium series beginning with uranium-238 has atomic weights $4n+2$, and is described as the $4n+2$ series; the thorium series starting with thorium-232 is the $4n$; and the uranium–actinium series based on uranium-235 is the $4n+3$. The missing family, $4n+1$, turned up when artificial isotopes were produced; it starts with plutonium-241 or the longer-lived neptunium-237, and ends with bismuth-209. None of its members are stable enough to have survived in the same way as uranium and thorium have done. Branch chains in the natural decay series filled the three remaining gaps in the periodic table: astatine (85), the last of the halogens, the alkali metal francium (87), and protoactinium

(91). The members of the series were originally called uranium-X, thorium-A (UX, ThA), and similar names; though these historical symbols do not indicate the chemical properties of the nuclide, they are sometimes still used if it is desired to draw attention to the position of the nuclide in the decay series.

The fact that all the natural decay series end with lead led to the realisation that elements are not necessarily of constant atomic weight, for lead extracted from a uranium-rich rock will contain slightly more lead-206 and lead-207 than lead from a thorium mineral, which will contain slightly more lead-208 and will therefore have a slightly higher atomic weight.

The *Group Displacement Law* emerged from the work on the decay series: loss of a particular particle always results in the same nuclear change, so that it is possible to work out without difficulty what an element will transmute into when it decays. The *Geiger–Nuttall* relationship was found empirically to apply to the alpha emitters: in any one series, the rate of decay is directly proportional to the energy of the alpha particle. The relationship is not perfectly linear, and tails off slightly with increasing alpha energy. The lines for the four series are parallel.

An experiment by Ramsay and Soddy in 1903, which was an unequivocal demonstration of the transmutation theory, classical in its elegance and simplicity, showed that helium had been formed from radium. The gas is in fact the alpha particles from the metal, which have picked up two electrons to balance their double positive charge. Dewar (1908) calculated the volume of helium that radium would produce, assuming one alpha particle per disintegration, as

$$\frac{3{\cdot}7\times 10^{10}\times 60\times 60\times 24\times 365\times 22{\cdot}4\times 10^{6}}{6{\cdot}023\times 10^{23}} = 43{\cdot}5 \text{ mm}^3 \text{ per g of Ra per year}$$

Four times this volume was actually found because at equilibrium radium gives, in effect, four alpha particles per disintegration. An atom of uranium produces eight atoms of helium in decaying to lead, and the calculated rate of production is 1 ml He per g of U in $8{\cdot}4\times 10^{6}$ years.

In 1903, Crookes observed that alpha particles produced minute flashes of light when they struck a screen of zinc sulphide. He made an apparatus based on the phenomenon, naming it a spinthariscope after the Greek word for scintillations. Crookes suggested that the spinthariscope could be used for detection and counting of particles, although Rutherford thought that only a few of the alpha particles incident on the screen produced scintillations, and that a determination of their number would have no physical significance.

For once Rutherford was wrong, as it was found later that the number of particles detected in this way and by gas ionisation were about the same, and scintillation counting was therefore a feasible method for detection and counting of radiation. At first, the only means for recording scintillations were the human eye and the hand tally counter, which restricted the method severely. Not more than two or three scintillations per second at most could be counted, so that only samples of comparatively low activity could be used. In consequence long drawn-out periods of unbroken concentration were required, and the counting of more than a few samples became a tedious chore, made worse by the fact that the scintillations are very faint and only just above the visual threshold of the unassisted eyes, which must be fully dark-adapted if anything is to be seen. Spinthariscopes can be had today for demonstration of scintillations, and they also serve as a salutary reminder of the difficulties that had to be coped with by the pioneers and which are not perhaps easy to appreciate fully in these more sophisticated days. In view of the difficulty of detecting the scintillations, it is hardly surprising that, as the ionisation chamber developed in reliability and sensitivity, scintillation counting was abandoned and for many years remained only of academic interest. The *photo-electric effect*, the conversion of light to electricity, had been known for some years; discovered by Hertz in 1887, and developed by Elster and Geitel, who made the first photo-electric photometer in 1892, it offered the possibility of replacing the human by the electric eye for detecting scintillations. But its exploitation had to wait until the 1940s when wartime research developed *photomultipliers* that could take the minute pulse of electricity from a scintillation and amplify it sufficiently for it to be counted. Scintillation counting then came back from the wilderness and expanded rapidly, so that it is now the preferred method of gamma counting, and a valuable alternative to gas ionisation for alphas and low-energy betas.

Radio-terminology

As the study of radioactivity developed, so a new terminology grew up with it. The first new name was coined by Soddy in 1913 to describe two atoms which are chemically the same, i.e. which have the same atomic number, but which have different atomic weights (mass numbers) and radioactive properties: as such atoms occupy the same place in the Periodic Table, Soddy used the word 'isotope', from the Greek for 'equal place'. The concept of isotopes solved the problem of why some elements had a fractional atomic weight: for example, chlorine is composed of two stable isotopes of atomic weights 35 and 37 in the ratio 3 : 1, giving the observed atomic weight of 35·5. The term nuclide describes a distinct nuclear species, i.e. an atom characterised by its mass number, atomic number, and nuclear energy state. Radionuclides are radioactive nuclides, isotopes are nuclides of the same element, and radio-isotopes are radioactive isotopes; radio-isotope and radionuclide are usually used synonymously. A parent radionuclide decays into a daughter, and a stable isotope is one which is not radioactive. Nuclear isomers are different energy states of the same nucleus which are capable of existing for a measurable time; the higher energy states are isomeric or metastable, and the lowest is the ground state. Nuclides of the same mass number are isobars, and those with the same number of neutrons are isotones. Radio-activity is the property of being radioactive, i.e. the spontaneous transmutation and emission of radiation by certain nuclides: activity is used as a quantitative expression of how radioactive a substance is, and should not be confused with the chemical meaning of effective concentration. The radioactive decay constant and half-life (*see* Section 2.2.1) express the speed at which a radio-nuclide decays. Protons and neutrons are collectively called nu-cleons; mesons are particles with mass intermediate between nucle-ons and electrons; and photons are quanta of electromagnetic radiation.

The unit of radioactivity originally chosen was the disintegration rate of one gram of radium, and the *curie* (abbreviated Ci) was defined as the quantity of an isotope that had the same disinte-gration rate as a gram of radium-226. Its size depended on the experimentally determined value of the radioactive decay constant (or of the half-life), and it varied with the accuracy with which these parameters were known: it was not therefore constant, but was a changing standard. This was obviously unsatisfactory, and a new definition was adopted in 1950, fixing the curie as that amount of any isotope giving $3{\cdot}7 \times 10^{10}$ disintegrations per second.

The measurement of radiation in terms of its biological effects was not (and is not) so easy, and the multiplicity of dosimetric units used over the years reflects this difficulty. An early attempt to quantify radiation in biological terms was the *skin erythema dose* (SED), the amount of radiation needed to redden the skin. It had more disadvantages than value, for it meant exposing hu-mans to radiation, the response took time to show (1–3 weeks) and was highly variable according to skin colour and nature, and it was too large to use with small doses. The International Commis-sion on Radiological Units (ICRU) redefined some units in 1962 and made several recommendations which have clarified the situation and are now generally accepted.

It is the custom in radiation work nowadays to refer to anything active as being *hot*, even though it may be stored in the deep freeze, and this usage of the word probably stems from the fact that these early-discovered radioactive elements are literally hot stuff: polonium is estimated to release 27·5 cal per Ci per d (1 cal = 4·2 J), and if $\frac{1}{2}$ g is placed in a capsule its temperature will rise to 500°C; polonium has been proposed as a light-weight heat source for spaceflight. Radium, too, is warmer than its surroundings. It is thought that the heat production from natural radioactivity helps towards keeping the Earth warm.

Most *natural radioactivity* is found among elements of high atomic number at the top end of the Periodic Table, with a few of the lighter elements also being radioactive. Those with long half-lives are sometimes known as primordial isotopes, e.g. uranium, thorium, rubidium-87, and some of the rare earths. Biologically, the most important is potassium-40, whose natural abundance is 0·0118% of all potassium, and which gives to every gram of potassium an activity of $8{\cdot}3\times10^{-10}$ Ci. It makes all living things radioactive, the quantity in an 'average man' producing about a million disintegrations per minute. Its decay product is argon, which accounts for the atmospheric concentration of this 'inert' gas (0·93%) being some three orders of magnitude greater than that of the others. Some light elements that are naturally radioactive have short half-lives, and this raises the question of how they were formed. Unlike primordial nuclides, they could not have been created when the Earth was formed aeons ago, or they would have had to be in impossibly large amounts originally; neither are they formed in the natural decay series. The answer lies in their constant formation through the agency of cosmic radiation, which acts on the nitrogen of the atmosphere to make carbon-14 and tritium, and on the chlorine of surface rocks to make a small amount of chlorine-36.

Transmutation of the base metals into gold was the dream of the ancient alchemists, and they sought endlessly for the 'philosophers' stone' that would accomplish this at a touch. Such a thing had long ago been shown to be impossible yet, for all that, transmutation was now revealed as a natural phenomenon which had been going on all the time unknown. Then in 1919, Rutherford discovered that a stable element could change into another when bombarded with nuclear particles, and accomplished the first *artificial transmutation*, of nitrogen-14 to oxygen-17, by bombarding it with alphas. In 1934 Curie and Joliot first produced artificial radio-isotopes (oxygen-17 is stable) by making phosphorus-30 out of aluminium-27. The development of accelerators (cyclotrons and the like) to speed up the bombarding particles made the work much easier, increasing the efficiency of the reactions and the yields of product elements, and prepared the way for the large-scale production of artificial isotopes that is carried out today. It was greatly helped following the research carried out under the stimulus of the 1939 war. About 1500 isotopes are now known, the vast majority being artificial.

The biological interest of radio-isotopes arose at first from their toxic properties, for the heavy elements do not occur naturally in living things and are in fact poisons. Some are highly unpleasant, such as polonium: weight for weight, it is some 10^{11} times as toxic as cyanide, one of the most toxic substances in common knowledge. The maximum permissible body burden is 0·03 μCi, equivalent to the tiny weight of $6{\cdot}8 \times 10^{-6}$ μg, and the maximum allowable concentration in air is 2×10^{-11} μCi/ml. For radium, the maximum body burden is 0·1 μg. Radon is dangerous because it can be inhaled and the lungs will be irradiated both with its radiation and that from the active material it deposits. Its maximum allowable concentration in air is 10^{-8} μCi/ml. Following the discovery of artificial transmutation, biologically useful isotopes became available, and they are now indispensable as research tools in every branch of biology and medicine.

By contrast, x-rays found biological application right from the start. Radiography soon became very useful in the diagnosis of internal defects like dislocations and broken bones, and radiotherapy was developed for the removal of warts and similar unwanted growths. But there was an entry on the debit side of the balance sheet, in the form of biological damage. The early workers with radiation did not at first realise the danger of what they were handling, though an awakening was not long in coming. Within four months of Röntgen's announcement of x-rays, a report appeared of hair loss following irradiation of the skull.

Radio-isotopes too were not without hazard, as Becquerel himself found out: he used to carry a phial of radium in his breast pocket to show off to his admiring friends, and soon became aware of the biological consequences by the effect on himself, as a burn developed on the skin below. Luminisers, applying luminous paint to watches and clocks, were in the habit of shaping their brushes to a fine point between their lips. The radioactivity in the paint —from radium—gave many of them fatal mouth and tongue cancers; they had radon in their breath, and radium was found in their bones after they died.

Over the succeeding years it was gradually realised, more and more clearly, just how unpleasant ionising radiations could be, and how important it was to use and handle them carefully and with respect for their power. This realisation came too late for many of the pioneer radiologists, and they suffered terribly from the effects of their diagnosing and healing rays. Others chose to close their eyes to it, putting the development of radiation before their own bodily health. They paid a heavy price, in pain, cancer, disfigurement, and even death: by 1936 more than 110 radiologists had lost their lives through radiation injury. There is indeed good cause for a profound sense of thankfulness that full appreciation of the vital necessity for safe handling of radiation came before the large-scale development of atomic power, or there might well have been many more tragedies. Though it must be remembered that the early workers just did not understand the power of radiation and did not take the precautions that are now known to be necessary, it is certain that many of them suffered more and died earlier than they need have done.

REFERENCES AND FURTHER READING

The Discovery of Radioactivity and Transmutation, Edited ROMER, A., (1963). Vol. 2 in Classics of Science Series, Dover Publications, New York, 233 pp.

SKLODOWSKA CURIE, M., (1898). *C.r. hebd. Séanc. Acad. Sci., Paris*, **126**, 1101–1103.

CURIE, P., and CURIE, S., (1898). *C.r. hebd. Séanc. Acad. Sci., Paris*, **127**, 175–178.

CURIE, P., CURIE, Mme P., and BEMONT, G., (1898). *C.r. hebd. Séanc. Acad. Sci., Paris*, **127**, 1215–1218.

Chapter 2

PHYSICAL ASPECTS OF RADIATION, OR RADIO-PHYSICS

2.1 INTRODUCTORY

2.1.1 Nuclear structure

The atomic nucleus can be considered to be made up of protons and neutrons. The number of protons determines the *atomic number Z*; it is the same as the number of orbital electrons, so that the atom as a whole is uncharged. The number of neutrons N added to Z gives the *mass number A* which, for all practical purposes, is equal to the atomic weight. The internationally agreed convention for describing isotopes is to place A as a superscript and Z as a subscript immediately before the chemical symbol: thus $^{14}_{6}C$ represents the isotope of atomic number 6, carbon, and atomic weight (or mass number) 14. Since the atomic number is unique to each chemical element, it is commonly omitted; ^{14}C conveys just as much information as the preceding representation. The inclusion of Z is helpful chiefly in the balancing of equations for nuclear reactions. A slightly different convention, used mainly in the United States of America, places A as a superscript following the chemical symbol, but has the disadvantage that this position is where the charge of an ion is written: thus the description of the alpha particle, a doubly charged helium nucleus containing 2 protons and of mass 4, is easy according to the first convention ($^{4}_{2}He^{2+}$) but becomes awkward with the second. When the name of an element is written out in full, it is followed by a hyphen and the mass number of the isotope, thus: carbon-14.

A pictorial or graphical representation of nuclear structure can be obtained by plotting the neutron number N for all known isotopes against the corresponding Z (Figure 2.1). This at once clearly shows that the make-up of the atomic nucleus is not random, since the graph takes the form of a fairly narrow band, gradually curving towards the axis on which N is plotted. The stable, non-radioactive isotopes lie in an even narrower zone in the centre of the band. It is therefore apparent that the ratio of N to Z, the ratio of

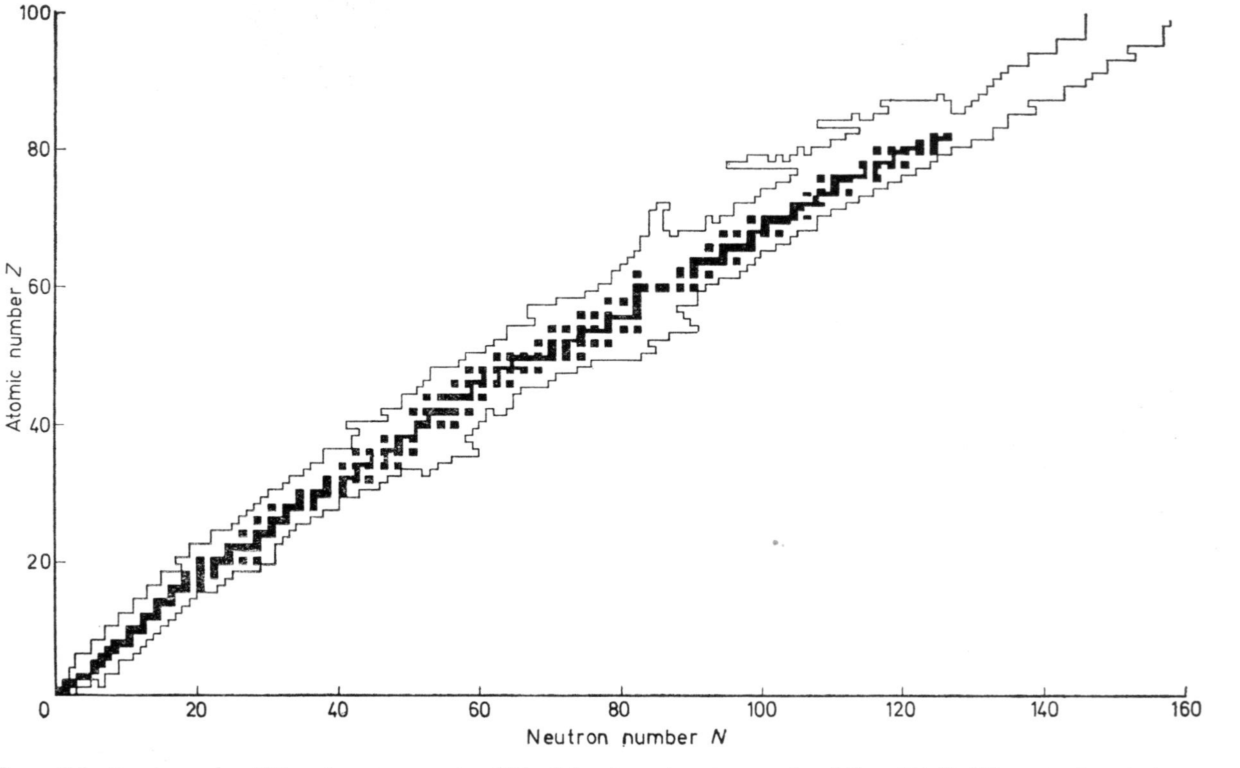

Figure 2.1. Atomic number (Z) and neutron number (N) of the elements; mass number (A) = N + Z. (The outer boundaries are extended from time to time as new radionuclides are discovered). ■ *Stable isotopes,* □ *Radioactive isotopes*

neutrons to protons (N:P ratio) must be within very narrow limits for the nucleus to be stable: for several elements it can have only one value, and there is in consequence only one stable isotope of such elements. The N:P ratio is 1:1 for the lighter elements, and alters gradually as Z increases until it reaches about 1·5:1 for the heaviest ones. Presumably if large numbers of protons were to be too close together, their mutual repulsion would be so great that the nucleus could not hold together. The more than proportionate increase in N with P is needed to keep the protons apart.

A nucleus whose N:P ratio is outside the limits for stability adjusts it in the appropriate direction, by a nuclear reaction that manifests itself as the emission of radiation, radioactivity. The half-life (Section 2.2.1) of such a nucleus and the energy of its radiation reflect its instability: in general, as the N:P ratio becomes further removed from the stability value, the more urgent is the need to adjust it, the more unstable the nucleus becomes and the shorter is its half-life. After a certain point, the nucleus is so

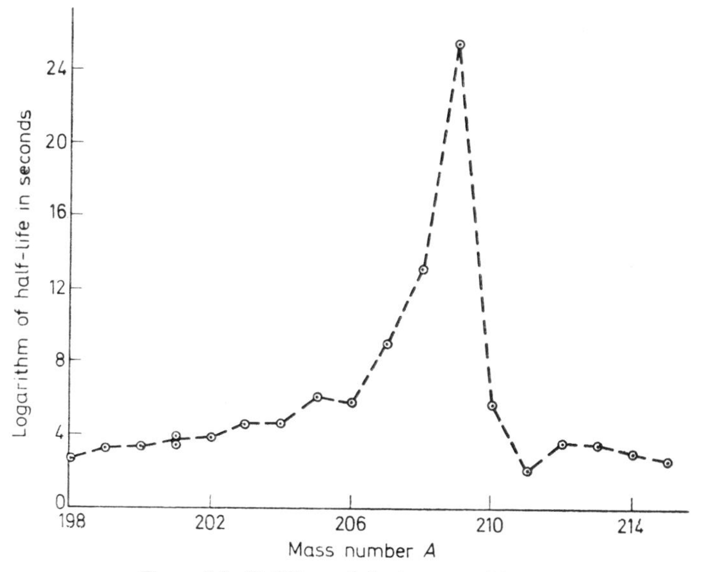

Figure 2.2. Half-lives of the isotopes of bismuth

unstable that it cannot exist at all: a good example is seen with the 18 isotopes of bismuth, shown diagrammatically in Figure 2.2. Bismuth was thought until recently to be the highest atomic number element to have a stable isotope, but it has now been displaced by the one below, lead (Z = 82), autoradiography in a nuclear emulsion (Section 3.4.4) having shown that bismuth-209 is in fact radioactive, with a half-life of over 10^{18} years. All elements with higher atomic number are radioactive, and the highest to occur naturally is uranium ($Z = 92$). An odd fact may be noted in passing, that elements of even atomic number often have more isotopes, both stable and radioactive, than those whose atomic number is odd. Charts of the nuclides, whose layout is similar to that of Figure 2.1, may be had; they contain much useful information in a handy form.

2.1.2 Ionising radiation

Ionising radiation is high-energy radiation that interacts with matter principally by ionisation, and it is on this criterion distinguished from other types of radiation that can damage living organisms, such as infra-red, laser, microwave, and sound. Ionising radiations of higher and lower energy are called *hard* and *soft* respectively. There are two types, particulate and electromagnetic.

Particulate radiations

These are fast-moving particles, charged or uncharged. They include the alpha and beta radiations, neutrons (Section 2.4), and heavy ions produced in particle accelerators, such as protons, deuterons, etc., which behave similarly to alpha particles, and will not be mentioned further. The velocity, energy, and mass of any particulate radiation are inter-related. The faster a particle travels, the more energy it has and, for a constant velocity, the energy is directly proportional to the mass. In other words, a heavy particle moves more slowly than a light one with the same energy.

Electromagnetic radiation

Electromagnetic radiation travels at the speed of light. Quantum theory shows that, while it is essentially a wave motion, it also has something of the nature of a particulate radiation in that the waves are in fact made up of discrete amounts (quanta) of energy, or photons, which are not further divisible. Its wavelength (or frequency) and energy content are related through Planck's constant; in practical terms, energy (in electron-volts*) = 12 400/

* One electron volt eV is the energy acquired by one electron in accelerating through a potential difference of one volt. It is equal to $1{\cdot}6\times10^{-12}$ erg, $1{\cdot}6\times10^{-19}$ J, or $3{\cdot}85\times10^{-20}$ cal. 10^6 eV = 10^3 keV (thousand or kilo-eV) = 1 MeV (million or mega-eV). 1 atomic mass unit (AMU) = 931 MeV.

wavelength (in Ångstroms). A photon of 1 eV is associated with a wavelength of 1·2395 μm. Thus ionising radiation in the electromagnetic spectrum is found only at the high-frequency end, in the region beyond the far ultra-violet of wavelength less than about 10^{-3} μm. Under certain circumstances it is possible for lower energy radiations, ultra-violet or even visible, to produce ions, as in the photo-electric effect, but they do not do so in living material; their interaction with matter is principally by excitation. They are not thought of in the present context as ionising radiations; nor are infra-red radiations, whose interaction is by mechanical vibration manifesting itself as heat.

2.2 RADIOACTIVE DECAY

2.2.1 Properties

Radioactive decay of an atomic nucleus is a completely *random* process, so that in any work involving radioactivity, either theoretical consideration or practical application, statistical methods have to be used. This fact may easily be forgotten and it is as well to bear it in mind when interpreting the results of an experiment for, unless correct methods are used, the conclusions drawn can be quite unjustified, or even false. However, one should not go to the other extreme, for although a lot of statistics may be a mathematician's dream, they can be a nightmare to the non-mathematical biologist. In any case, too, no amount of statistical dressing up can possibly make poor results significant when in fact they are not.

The basic concept of radioactive decay is that there is a constant probability that an unstable atom will decay within a given period of time, this probability being unaffected by the history of the atom, its chemical or physical state, or by the passage of time. Thus the rate at which a quantity of isotope decays, i.e. its activity, is proportional to the number of unstable atoms present: as the unstable atoms break up, their total number decreases, and so the graph of the activity against time is a typical exponential curve, flattening out gradually to run increasingly parallel to the time axis (Figure 2.3a). According to the graph, there should always be some activity present, since the curve meets the time axis only at infinity. However, the statistical theory is based on random decay among a large number of radioactive atoms, and does not hold for small numbers; in addition, the supply of isotope is not infinite, and so the activity can be regarded as eventually falling to zero.

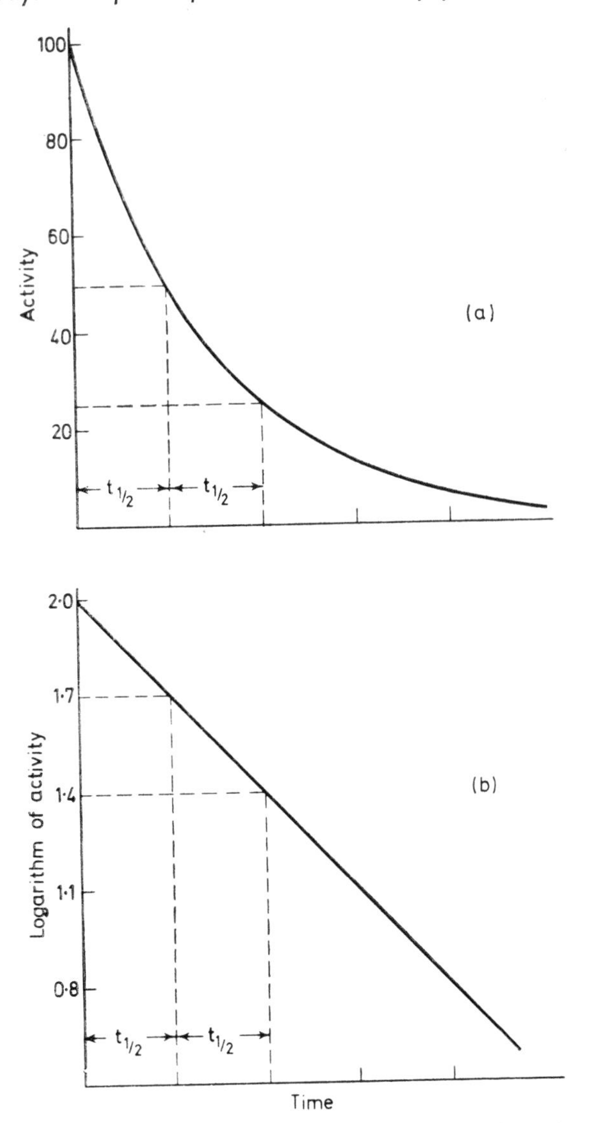

Figure 2.3. Radioactive decay: (a) linear scale, and (b) semi-logarithmic scale

The exponential curve of radioactive decay is mathematically expressed by the equation

$$\frac{dN}{dT} = -\lambda N \tag{2.1}$$

i.e., the rate of change of the number of radioactive atoms is proportional to the number present (N), the two being related by the radioactive *decay constant* (Greek lambda). The minus sign is introduced because N is decreasing. The value of λ is characteristic for each isotope, and is, as mentioned above, truly constant. It is defined as the fraction of a quantity of isotope decaying in unit time, and it has the units $(\text{time})^{-1}$.

Straight-line graphs are easier to handle than curves, and an exponential relationship yields a straight line if it is plotted semi-logarithmically: the independent variable x (in this case time) on a linear scale against the dependent variable y (activity) on a log scale (Figure 2.3b). This is expressed mathematically by integrating Equation 2.1 and re-arranging it into the following form:

$$\log_e \frac{N_t}{N_0} = -\lambda t \qquad (2.2)$$

or

$$\log_{10} N_t = \log_{10} N_0 - 0{\cdot}4343\lambda t \qquad (2.3)$$

$\log_e = \ln$, 'natural' or 'hyperbolic' logarithm; $\log_{10}$ or 'common' logarithm $= \log_e/2{\cdot}303$; N_0 = number of radioactive atoms, or the activity, at time zero; N_t = activity at time t.

Since the rate of decay of a radioactive element is exponential, falling off with time, it is meaningless to speak of the 'life', meaning the whole life, of the element, since this always tends towards infinity. A more meaningful expression of how long an isotope lasts is provided by two parameters, the half-life and the average life.

Half-life

Expressed as $t_{1/2}$, the half-life is an extremely useful quantity, for it has a simple and constant inverse relationship to the decay constant, and is therefore an important diagnostic property of an isotope. Indeed, the first step in the identification of an unknown isotope is often a determination of its half-life. Half-life is defined as the time taken for the activity of a quantity of isotope to decay to, or by, one half ($t_{1/2}$ in Figure 2.3). It is perhaps worth making it quite clear that the activity of a quantity of isotope is *not* reduced to zero after two half-lives, but to a quarter of its original value. This mistake can easily be made if the definition of half-life is not taken along with a clear understanding of the nature of decay.

In Equation 2.2 above, if N_t is made equal to half N_0, then t will be $t_{1/2}$, by definition, and the equation becomes

$$\log_e \tfrac{1}{2} = -\lambda t_{1/2}, \quad \text{or} \quad 2{\cdot}303 \log_{10}(\tfrac{1}{2}) = -\lambda t_{1/2}$$

whence
$$t_{1/2} = \frac{0{\cdot}693}{\lambda} \tag{2.4}$$

Half-lives vary over a tremendously wide range, from some 10^{15} years to less than 10^{-9} seconds, a range of over 10^{30}; no other physical parameter shows such a wide range of variation. Isotopes with very short half-lives are obviously no use for experimental work, and in fact most isotopes have half-lives too short to be of any practical value. There are about 200 nuclides with useful half-lives, of several minutes and upwards, and most of these are commercially available.

The precise determination of half-lives is difficult technically, although, if they are of the order of minutes or a few days, they can easily be estimated in the laboratory. As stated above, a semi-logarithmic plot of activity against time is a straight line, whose gradient is $-0{\cdot}4343\ \lambda$ (Equation 2.3), N_0 being constant. Substituting in Equation 2.4 gives the expression $t_{1/2} = -0{\cdot}3010/\text{gradient}$. The appearance of the term 0·3010, which is $\log_{10} 2$, is not due to chance, but follows directly from the definition of half-life, a decrease of $\frac{1}{2}$, or a factor of 2, being represented by 0·3010 on a logarithmic scale. Thus, a half-life is estimated by making a semi-logarithmic graph of the activity of a sample under constant conditions against time: the gradient is then substituted in the expression above, or alternatively the time taken for the activity to fall to $\frac{1}{2}$, $\frac{1}{4}$, etc., of its original value is read off the graph directly (Figure 2.4). It is advisable to take a large number of readings to

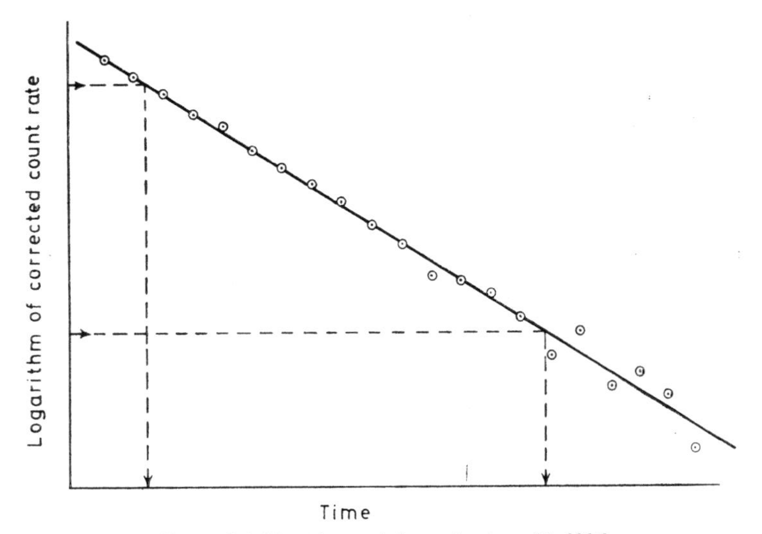

Figure 2.4. Experimental determination of half-life

construct the graph, rather than just two, in order to even out the inevitable statistical error in any one count; the experiment should also be extended over as long a period as possible. The tendency towards increased scatter of the results as the activity diminishes may be noted.

Short-lived isotopes are not easy to deal with because of this very fact (they have disappeared before it is possible to measure very much) and long-lived ones present the problem that the quantity present, and hence the observed activity, does not change significantly in years. Here the determination of lambda has to be undertaken, an exercise involving the determination of the abundance of radioactive atoms in the sample by such means as mass spectrometry, or by autoradiography in a nuclear emulsion (Section 3.4.4).

Average life

The average life is the reciprocal of the decay constant. It is useful in assessing the total amount of radiation emitted in a given time, as in radiotherapy, where it is essential to know the total radiation dose delivered to the patient during the period of exposure. From Equation 2.4, it can easily be worked out that the average life = $1{\cdot}443\ t_{1/2}$.

2.2.2 Consequences of decay

The fact that radioactivity decreases with time has three important consequences.

First, in radiotracer experiments, *two corrections* for decay are often required. Where the half-life of the isotope is commensurate with the duration of the experiment, the measured activities of samples taken during the experiment have to be corrected so as to be comparable with the amount of activity that was put into the system initially. They are calculated back to what they would have been at a fixed reference point in time, which is usually taken as the beginning of the experiment. Equations 2.3 and 2.4 can be combined to yield:

$$\log_{10}N_0 = \log_{10}N_t + \frac{0{\cdot}3010t}{t_{1/2}} \qquad (2.5)$$

N_t is the observed count, and N_0 is the desired corrected count; t is the time between the reference point and the time of counting the sample. *See* p. 278 for worked examples.

Another correction involving decay is the calculation of how much activity is left in a sample after some time has gone by, as for example to see if there is enough activity left for another experiment, or whether the activity of a batch of waste has declined sufficiently for it to be thrown away safely. If in Equation 2.5, N_0 is taken as 100, then N_t will be the percentage activity left after time t, and is given by:

$$\log_{10} N_t = 2 - \frac{0{\cdot}3010t}{t_{1/2}}$$

If a particular isotope is being used routinely, it is easier and more convenient to draw up a decay table rather than to use the formulae each time to calculate the corrections. Such a table will show how much of the original activity is left as time goes on, and/or the factors by which observed activities must be multiplied to correct them back to zero time. The two sets of figures are the reciprocals of each other. The table may be specific for one particular isotope, so that each isotope in use has to have its own table. If the time scale is in half-lives rather than in conventional time units (hours and days) one table can be used for any isotope, although each observed time has then to be converted into terms of half-lives. If only approximate corrections are required, they may be read off from a decay graph, a straight line drawn through two calculated points on semi-log paper.

Secondly, the *biological effect* of an isotope, all other things being equal, will be less the faster the isotope decays, since obviously it will be in the organism for a shorter time (Section 6.1.4).

Thirdly, the practical problem of *disposal of radioactive waste* is greatly simplified if the isotope decays away quickly. No special precautions in disposal may then be necessary: it may simply be sufficient to store the waste until its activity has decayed away sufficiently (Section 6.6.8).

2.2.3 Composite decay

If two isotopes are present together in one sample, the observed decay graph, activity versus time, will be a composite of the two individual ones. When the half-lives are sufficiently dissimilar, it has two parts (Figure 2.5). It starts as a curve (C) which gradually flattens out as the shorter-lived isotope decays away: when it has effectively all gone, the curve turns into a straight line which is the characteristic decay line for the longer-lived isotope (A). If this line is extrapolated back towards zero time and subtracted

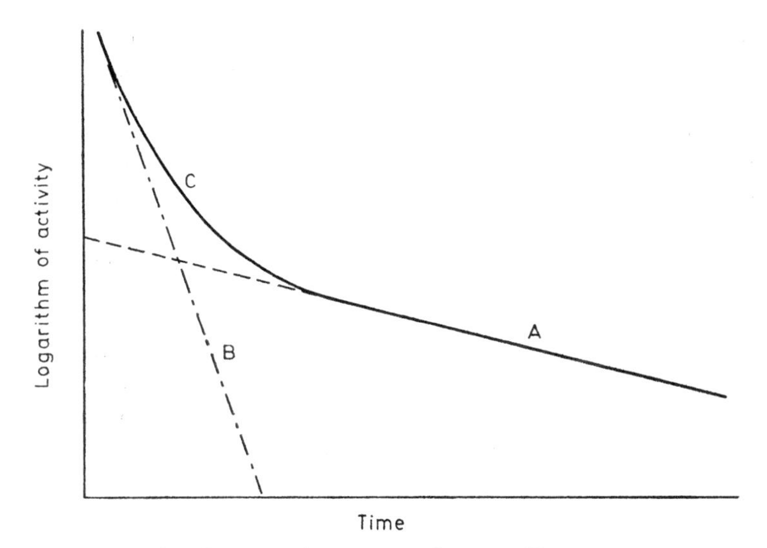

Figure 2.5. Composite decay. For explanation of lettering, see text

from the observed curve, the decay line of the shorter-lived isotope can be derived (B). The half-lives of the two are then easily calculated, and two isotopes in admixture may thus be *identified.* The difficulty of doing so increases as the half-lives approach each other, and as the quantity of the longer-lived one increases relative to that of the shorter-lived. Theoretically, a mixture of many isotopes could be analysed in this way, but in practice not more than three can be dealt with. It is essential, of course, for one isotope to decay away completely and for both half-lives to be fairly short, otherwise part A of the curve will be parallel to the time axis and the corresponding half-life will be unobtainable. This restricts the usefulness of the method.

When the half-lives and identities of the isotopes are known, their *quantitative determination* is simple, provided that the half-life of one is short compared with that of the other. It is carried out by counting the sample before and after a suitable period of time, so that two simultaneous equations are obtained. If x is the initial count rate of isotope X and y that of isotope Y, then:

$$x+y = \text{observed count 1}$$

and

$$ax+by = \text{observed count 2}$$

The coefficients a and b represent the fractions of X and Y respectively which have decayed in the time between the two observed counts. Knowing this time and the half-lives of X and Y, the values of *a* and *b* may be calculated and substituted in the equations, which are then solved for *x* and *y*.

Without going into detailed mathematics, it is obvious that *a* and *b* must be different, and sufficiently different to give statistical validity, and the time that must elapse to make them different must be within the time scale of the experiment, e.g. it is no use trying to count sodium-22 ($t_{1/2} = 2 \cdot 6$ years) and chlorine-36 ($t_{1/2} = 300{,}000$ years) together by this means in practice, even though it is theoretically satisfactory: one experiment could last several years! A special case, the easiest to handle, is when the shorter-lived isotope is able to decay away completely, *a* or *b* then being zero: here, the second count is due entirely to the longer-lived one, and it is simply corrected back to zero time (if necessary) and subtracted from the first count. (Example: sodium-22 ($t_{1/2}=2{\cdot}6$ years) and sodium-24 ($t_{1/2} = 15$ hours) together, second count a week after the first.)

2.3 UNITS OF RADIOACTIVITY

2.3.1 The Curie

The curie (abbreviated Ci) is defined as the quantity of an isotope which gives $3{\cdot}7\times10^{10}$ disintegrations per second. It is approximately equal to the activity of 1 g of pure radium-226.

1 g of ^{226}Ra contains $6{\cdot}023\times10^{23}/226$ atoms, and $\lambda = 0{\cdot}693/1622$ years (present best estimate).

Thus the number of atoms disintegrating per second

$$= \frac{0{\cdot}693\times6{\cdot}023\times10^{23}}{1622\times365\times24\times60\times60\times226} = 3{\cdot}61\times10^{10}$$

Two points must be noted.

1. The curie refers to the number of disintegrations actually occurring in a sample, not merely to the number of 'counts' picked up by a detecting device, whose efficiency is never 100%; a correction must be applied when determining the curie strength of a source from the experimentally determined disintegration rate or count rate.
2. The curie is based on the number of disintegrations and not on the number of particles emitted. One curie of, e.g. phosphorus-32, which gives off one beta particle per disintegration, will emit a different amount of radiation from one curie of

cobalt-60, which gives off one beta particle and two gamma rays. This has to be taken into account when calculating the disintegration rate from the observed counting rate, and also in dosimetry, when calculating how much radiation is delivered by a given amount of isotope in a given time.

It is self-evident that the weight of a curie varies with the half-life and with the atomic weight of the isotope.

$$\text{From Equation 2.1, } N = \frac{-\mathrm{d}N/\mathrm{d}t}{\lambda}$$

$$\therefore\ N = \frac{3{\cdot}7\times10^{10}}{\lambda} \text{ atoms/Ci}$$

$$\therefore \text{ Weight of 1 Ci} = \frac{3{\cdot}7\times10^{10}}{\lambda}\times\frac{A}{6{\cdot}023\times10^{23}} \text{ g}$$

where A is the atomic weight of the isotope, since this amount contains Avogadro's number of atoms.

$$\therefore \text{ Weight of 1 Ci} = \frac{3{\cdot}7\times10^{10}}{0{\cdot}693\times6{\cdot}023\times10^{23}}\times A\times t_{1/2} \text{ g}$$

where $t_{1/2}$ is the half-life expressed in seconds.

If $t_{1/2}$ is in days, the weight of 1 Ci becomes $7{\cdot}66\times10^{-9}\times A\times t_{1/2}$ g.

The figure obtained for the weight of 1 Ci in this way will obviously depend on the accuracy with which the half-life is known, e.g. the half-life of carbon-14 is quoted variously as between 5568 and 5760 years, giving from 0·218 to 0·225 g as the weight of one curie.

It may also be noted that, if n atoms decay per unit time in a sample of isotope, then the whole amount of the sample contains n/λ atoms of the isotope.

Specimen values for the weight of 1 Ci are shown in Table 2.1.

Table 2.1. WEIGHTS OF 1 Ci AND HALF-LIVES OF VARIOUS ISOTOPES

Isotope	$t_{1/2}$ (a = *year*)	W(g)
^{238}U	$4{\cdot}51\times10^{9}$ a	$2{\cdot}99\times10^{6}$
^{14}C	5730 a	0·218
^{35}S	87·2 d	2×10^{-5}
^{131}I	8·04 d	$8{\cdot}08\times10^{-6}$

Abbreviation for the curie

This abbreviation is Ci, having been changed fairly recently from c for some reason, perhaps that the abbreviation for a picocurie (pc) could be confused with per cent; the small c is still frequently found in the literature, and a capital C was sometimes used also. The curie is a large unit for tracer work and the standard subdivisions milli- and micro- are commonly used (mCi and μCi). Where activities are lower still, as in work on environmental contamination with radionuclides, nano-, pico-, and occasionally femto-curies (nCi, pCi, fCi) are used, representing respectively 10^{-9}, 10^{-12}, and 10^{-15} of a curie.

2.3.2 Specific activity

The term specific activity expresses the relative abundance of the radioactive and the stable isotopes, giving the amount of activity per unit quantity of the sample. There are several ways in which it may be expressed, e.g. mCi/unit mass, μCi/m mol, counts per min per unit volume of a gas, ratio of active atoms to stable atoms, and so on. Which particular form is used depends on the requirements of the experiment and of the sample. Comparison of different samples is facilitated and made more meaningful when their activities are expressed as specific activities rather than as curie strengths. The highest possible specific activity is attained with a preparation of 100% isotopic abundance, in which all the atoms (initially at least) are active, with no stable nuclides of the element present. The value of the maximum specific activity can be calculated from the half-life of the isotope ($t_{1/2}$) and its atomic weight (A): it is $1{\cdot}3 \times 10^8/At_{1/2}$ Ci/g, if $t_{1/2}$ is expressed in d.

Such a preparation is correctly described as *carrier-free*, since there are no stable 'carrier' atoms of the element present. This condition is rarely attained in practice, however, and material of high specific activity is often loosely referred to as carrier-free, although it should more correctly be termed material of high isotopic abundance. The Radiochemical Centre, Amersham, defines a carrier-free preparation as being one to which no carrier has been added, and for which precautions have been taken to minimise contamination with other isotopes. The statement that an isotope is 'carrier-free' should always therefore be qualified with a statement of the actual specific activity.

Stable isotopes are quantified in terms of *atom per cent excess*

(APE), the proportion of isotope over and above that which is normally present naturally. The natural abundance must be subtracted from the abundance observed in an enriched (labelled) compound to find the APE.

2.4. PARTICULATE RADIATION

2.4.1 Alpha particles

Alpha radiation is typical of the natural decay series, i.e. of elements of higher atomic number, above $Z = 82$. Alpha emitters are not normally found in living organisms, but their accidental presence is very hazardous, mainly because of the radiological danger from internal emission of alpha particles and also because many high-Z elements are chemical poisons. An alpha particle is a helium nucleus, ${}^{4}_{2}He^{2+}$, so that alpha emission reduces the atomic weight of the nucleus by 4 and the atomic number by 2.

Alpha radiation is *mono-energetic*, i.e. all the alphas from a given nuclide are always emitted with one, or a few, discrete energies. This offers a means of identifying an unknown alpha emitter : if the energy (or energies) of its alpha particles can be determined, reference to the books should show what isotope is present. The energies of alphas are quite high, in the range 4 to 8 MeV. They are relatively slow-moving, their initial velocities being about 5–7·5% that of light.

2.4.2 Beta particles

Beta radiation is the most common type of emission from a radioactive nucleus. Beta particles are high-speed electrons, their initial velocity being about 33–99% that of light. There are two kinds, which differ in the sign of the charge they carry: negative electrons, or negatrons (β^-), carry unit negative charge, and positive electrons, or positrons (β^+), carry unit positive charge. Negatrons are very like cathode rays, which are streams of electrons.

Negatrons are emitted when there are excess neutrons in the nucleus : one neutron transmutes into a proton, which stays in the nucleus, and an electron, which is emitted. The atomic number therefore increases by one, with no significant change in mass, and the element moves one place up the periodic table. Positron emission is the reverse, and takes place if excess protons are present. One of them ejects a positron and turns into a neutron, which remains in the nucleus, so that the atomic number decreases by one and the element moves one place down the periodic table. Again, there is no change in mass. Negatron emission is found more commonly than positron emission, since many isotopes are manufactured by neutron irradiation of stable elements, which gives the product nuclei an excess of neutrons.

The properties of a positron exactly resemble those of a negatron, except that a positron has a transient existence. Immediately its kinetic energy has been expended, a positron is attracted to the nearest electron, and the two react: negative annihilates positive, and all of their mass is converted to energy, appearing as two gamma rays of 0·51 MeV each, emitted opposite ways along a straight line as '*back-to-back*' radiation.

An important point of difference from alpha radiation is that beta particles from a particular isotope have a *range of energies* up to a maximum (E_{max}) which is characteristic of the isotope. The maximum may be low, e.g. 0·018 MeV for tritium, or high, e.g. 4·81 MeV for chlorine-38. Only a few particles are emitted with maximum energy, and a more realistic measure of the effective energy of the radiation is given by the mean energy, E_{mean}, defined by saying that the number of particles with energy below it is equal to the number with energy above. A large, often the largest, proportion of particles have energies around this value, which is about a third of E_{max}. The distribution curves for beta emission, showing how many particles are emitted at each energy, follow a general pattern which is somewhat reminiscent of a normal distribution curve, centred approximately about E_{mean}. For soft beta emitters, only the right-hand side of the curve is seen, the axis of zero energy cutting off the left-hand end: as the radiation becomes harder, more and more of the curve appears. It should be noted, however, that the precise shape of the curve varies from one nuclide to another, for beta emitters are divided into a dozen 'transition types', a different nuclear transformation taking place in each. It may also be noted that the low-energy portion of each spectrum has to be calculated theoretically since experimental determination is not practicable (Figure 2.6).

The range of energies shown by the beta particles emitted from one nuclide disturbed the theoretical physicists and, to explain it, the theory was put forward by Pauli that each disintegration event in beta decay involves the same total energy release, equal to E_{max}, shared between the observed beta particle and a *neutrino* which carries off the rest of the energy. The proportion of the energy taken by the beta particle and the neutrino varies from one disintegration to another. The neutrino has no charge and zero (or possibly very small) mass: consequently it has virtually no interaction with matter. It was therefore exceedingly difficult to dem-

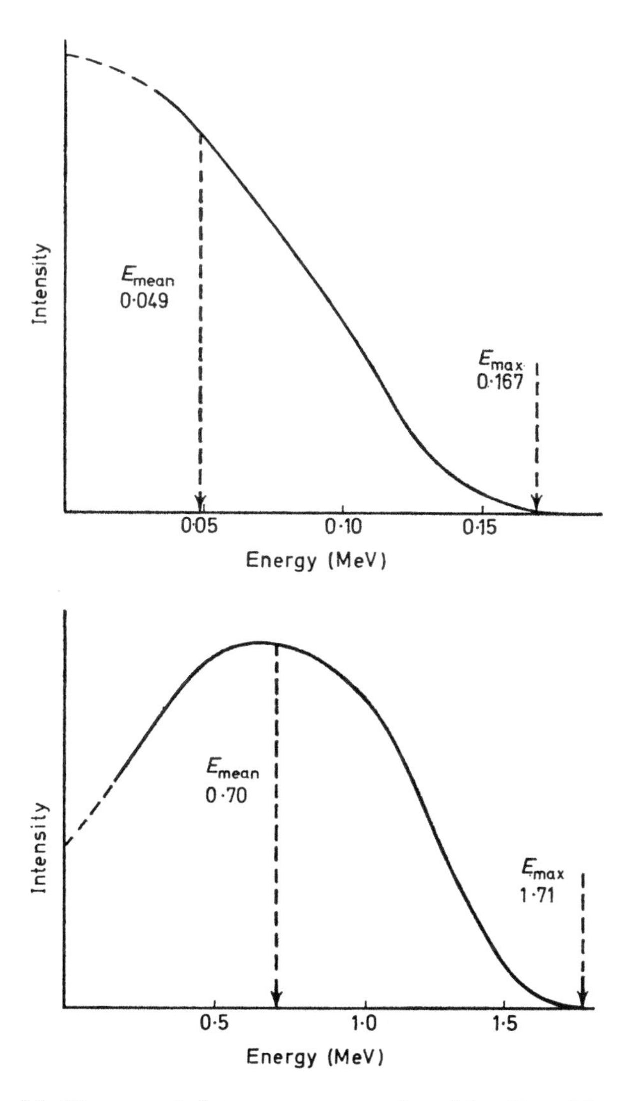

Figure 2.6. Diagrammatic beta spectra: upper, for sulphur-35, and lower, for phosphorus-32

onstrate that the neutrino actually existed, and in fact it remained theoretical for 25 years until an elaborate experiment in 1956 actually managed to detect one or two.

2.4.3 Interaction of alpha and beta particles with matter

Collisions between charged particles and atomic nuclei are rare, and their effects on matter arise virtually entirely from their interaction with the local electric fields of the atoms, by which the particles lose energy. Some of this energy appears as heat. In addition, *bremsstrahlung* may be produced (German for braking radiation—it is written in English with an initial small letter instead of a capital); this is electromagnetic radiation arising from the retardation of charged particles. It occurs mainly when energetic beta radiation interacts with high atomic number materials, either with the source itself or with surrounding materials, and it can be a nuisance in detecting the beta radiation that gave rise to it, although bremsstrahlung sources do have a number of uses. Bremsstrahlung is negligible with soft beta and low atomic number material, and also with alpha radiation. But more important from the radiobiologist's point of view is the transfer of energy from the particle to an orbital electron of the atom. One of two things may happen.

Excitation

In excitation the electron is raised to a higher energy state (or orbital): it then quickly drops back to its ground state and releases the energy it has acquired as a photon (electromagnetic radiation); this is of a fixed frequency, which depends on the energy relationships within the atom whose electron is excited. If the photons are in the visible range, they can be seen, and on this is based the detection of radiation by scintillation counting (Section 3.3).

Ionisation

Ionisation takes excitation a step further: the orbital electron acquires so much energy that it is completely removed from the atom, which consequently is left with a positive charge. An ion pair (negative electron + positive atom) is thus formed. The electron may leave its atom with such a speed (or energy) that it can itself interact with matter, and cause secondary excitations and ionisations: it is then known as a *delta-ray*, though in effect it is exactly the same as a beta particle or any fast moving electron.

Ionisation is the most important process in the transfer of energy from ionising particles to matter. The intensity of ionisation along the path of an ionising particle is determined by the rate at which it loses energy, or *linear energy transfer* (LET), which is conventionally expressed in energy units (keV) per unit of path length (μm). The term specific ionisation has been introduced to describe the intensity of ionisation in air, and it is defined as the number of ion pairs produced per centimetre path length in air at standard pressure. About 34 eV are needed to form an ion pair in a gas, so the specific ionisation may be easily calculated from the LET and vice versa. The energy needed to form an ion pair in liquids and solids is generally taken to be the same as in air. Thus a 1·7 MeV particle, such as an E_{max} beta from phosphorus-32, can produce some 5×10^4 ion pairs in air.

The rate at which a particle loses energy increases as the square of its charge: increasing charge increases the probability of interaction with the local electric field of an atom. It is also inversely proportional to the velocity: a slow moving particle will remain near an atom longer than a fast moving one, and will be more likely to interact with it. Thus alpha particles, with their double elementary charge and relatively low speed, have a much greater probability of interaction and are much better ionising agents than betas; the specific ionisations are of the order of 40 000/cm and 50–100/cm respectively, and they may be described as sparsely- and densely-ionising radiations. However, they both produce about the same number of ionisations per unit energy, alphas giving one ion pair per 30 eV dissipated on the average, and electrons one per 34 eV. As the particle slows down by its interaction with the matter through which it is passing, the probability of its interacting increases, so that the end of its path is marked by a sharp increase in specific ionisation, falling rapidly to zero when the particle has expended all its energy and effectively come to rest. This 'tail' is seen with beta particles when their energy is less than about 0·1 MeV—over a very short distance, ionisation is a few hundred times greater than it is initially—with alphas, the tail is at lower energy but longer. The variation in specific ionisation with distance may be plotted accurately for alpha particles, since they are mono-energetic and travel in approximately straight lines, but this cannot be done with betas.

Absorption

The range of a particle in a given material is the distance for which it will penetrate from the source, and is a measure of the ability of the material to stop the radiation, or its *stopping power*. The range of a particle is determined by the rate at which it interacts with the material and loses energy, and this depends essentially on the number of electrons per unit volume.

Alphas (and other heavy ions) have a high LET, and therefore a very short range, even in air. (A rough working guide is that the range of an alpha in cm of air is the same as its energy in MeV: the precise figure is slightly lower.) They are difficult to deflect, either by interaction with atoms or by electric or magnetic fields, simply because they have a high momentum, so that their paths are nearly straight lines and their range is nearly the same as the path length. The range is also clearly defined for an alpha particle of given energy. Although it is not strictly true to say that all the particles of a given energy are absorbed at one definite depth in a material, and that the number of particles detected at increasing distance from the course will be constant and then drop suddenly to zero, the decline in numbers is sufficiently steep to give only a short distance between the last point at which the intensity is 100% and that at which it is zero. The mean range is the distance from the source where the intensity is half the original.

The range of alpha particles in materials other than air is much smaller, and is measured in microns rather than centimetres. Such very small distances are difficult to measure in practice; they also vary widely, over several orders of magnitude if the range in air is included. However, the density of the material is highly important in determining the range, the range decreasing roughly with increasing density. The stopping powers of different materials may be made much more easily comparable by taking account of this and expressing the thickness needed to stop the radiation, not as an actual linear distance, but in terms of the weight of material per unit area, the *equivalent thickness*. This has the practical advantage that weight and area are more easily measured than very small distances. Equivalent thicknesses are usually given in milligrams per square centimetre (mg/cm^2); they can be converted to actual thicknesses simply by dividing by the density of the material.

Beta particles, unlike alphas, are easily deflected by passage near other atoms and by magnetic and electric fields, because they are so much lighter. Their paths are by no means straight, and their ranges are much less than the path lengths of the individual particles. However, the probability of their interacting with matter is less than with alphas, because of their greater speed and single charge, so that their ranges are greater than those of alphas although they have lower energy. The range of the beta particles from a given isotope is not well defined because of their continuous energy spectrum and also because of scattering. The absorption of beta particles by matter is the resultant of a number of separate processes, which purely by chance combine so as to give for most of the way an approximately linear relation between the logarithm of the activity detected and the thickness of material: though the relation is only pseudo-exponential and not truly so, it does mean that the exponential equations for gamma absorption can be applied empirically to beta radiation. The absorption curve levels out at the foot, as a result of bremsstrahlung.

The maximum range of beta radiation, which is necessary to determine E_{max} in the identification of an unknown isotope, is found accurately by *Feather analysis* of an absorption curve in aluminium and comparison with a beta emitter of known range. With a pure beta emitting isotope, the two parts of the absorption curve may be extrapolated, the point where they meet giving the visual range, which is less than the true maximum range. The maximum range is said to be the absorber thickness reducing the beta flux by a factor of 10 000. The equation $d_{1/2}=46\ (E_{max})^{3/2}$ is an approximation for the calculation of E_{max}, in MeV; $d_{1/2}$ is the thickness, in mg/cm^2, required to reduce the radiation intensity to a half.

The percentage of beta particles transmitted through a given thickness of absorbing material bears a sigmoidal relationship to the logarithm of their maximum energy (Figure 2.7). If the thickness is increased, the curve will move to the right, so that there is a whole family of such curves. This is of some practical importance

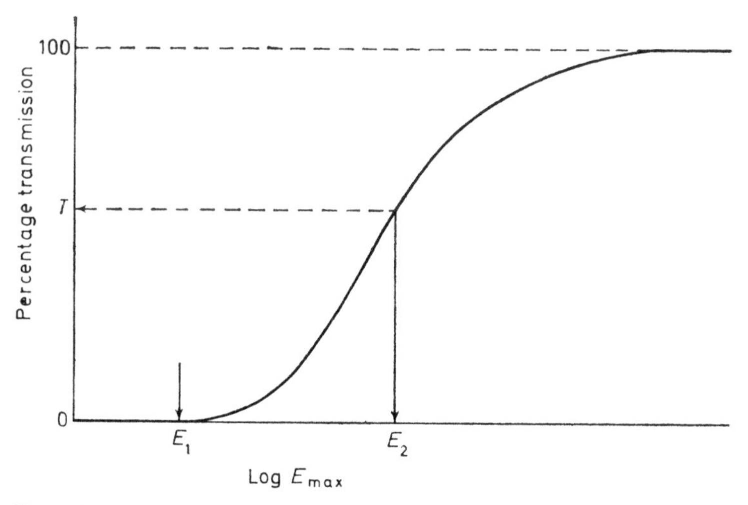

Figure 2.7. Transmission of beta particles through a given absorber. For explanation of lettering, see text

as the basis for quantitative analysis of a mixture of two beta emitters. An absorber is chosen of thickness sufficient to stop completely the softer radiation (energy E_1): knowing the transmission (T) of the harder radiation (energy E_2) through the absorber, the amounts of each isotope in the mixture can then be determined easily by reference to standards each containing one of the two isotopes. Absorption is also significant in the detection of low-energy beta particles by Geiger–Muller counting (Section 3.2.4).

Beta particles can be scattered so much that they are completely reversed in direction and travel back towards the source rather than away from it: this is the phenomenon of *backscatter*. In a somewhat oversimplified way, it can be regarded as the converse of external absorption in that the particles appear to be bouncing off the material instead of passing through it. Backscatter increases at first directly with increasing thickness of scattering material, from which it follows that scatter does not take place at the surface but within the thickness of the material. The increase is steeper the lower the beta energy. The backscatter graph then flattens out to a constant value: the thickness of material at this point is the saturation backscatter thickness for the particular radiation, which might be expected to be half the range, but is in fact about a fifth, since the particles are scattered so much within ihe material. The backscattering factor expresses the increase produced in the count rate from a source due to backscatter. It is independent of beta energy between 0·3 and 2·3 MeV and increases with atomic number of the scattering material, sharply at first and then less sharply.

Radiation may be absorbed not only externally, i.e. by material outside the source but internally, by the material of the source itself. This is *self-absorption*, and it applies both to alpha and beta particles, but it is of greatest consequence with low-energy beta emitters. As the thickness of a layer of such an isotope is increased, the observed count rate at first increases linearly with it, but soon starts to fall off, until finally the count rate is constant and independent of source thickness. The particles from the innermost part of the source are now being completely absorbed within the source material before they can reach the outside and enter the detector, and any further addition of source material merely absorbs the same amount of radiation as it emits. The source is now at *saturation* or *infinite thickness*, and the count rate is proportional to the specific, not to the total, activity. Self-absorption also occurs in liquids, and increases slightly with the specific gravity of the liquid.

It is very difficult to prepare reproducibly very thin samples in which self-absorption is negligible, and the easiest way round the difficulty is to eliminate self-absorption effects altogether and work with samples of saturation thickness. When this is not possible, a correction curve must be constructed by counting several sources, of the same specific activity and surface area and gradually increasing thickness, and plotting their apparent specific activity (count rate per milligram, best taken logarithmically) against their thickness (or weight). Extrapolation to zero thickness shows the true specific activity with no self-absorption: this is a hypothetical concept, requiring layers of sources so thin as to be unattainable in practice; it can be regarded as 'infinite thinness' of source material. A second graph is now constructed, taking the true specific activity as 100% and plotting the other values of specific activity as percentages of this against the corresponding source thickness. The extent of self-absorption in a moderately thick source, i.e. between zero and infinite thickness, is then read off from the curve, the apparent specific activity giving the correction factor to obtain the true specific activity from the observed result. Sources of different thicknesses may thus be compared. A correction curve is valid only for the isotope and for the counting assembly and geometry used in its construction.

The '*Radiochemical Manual*' (see references at end of Chapter 4) gives a nomogram of external and self-absorption of beta particles, relating their energy and the absorber thickness to the percentage absorption.

2.4.4 Neutrons

Neutrons are produced when alpha or gamma radiation interact with a light element such as beryllium:

$$^{9}_{4}\mathrm{Be} + ^{4}_{2}\alpha \longrightarrow ^{12}_{6}\mathrm{C} + ^{1}_{0}\mathrm{n}$$

This is described as a (α, n) reaction, and is written in shorthand notation as

$$^{9}_{4}\mathrm{Be}(\alpha, \mathrm{n})\ ^{12}_{6}\mathrm{C}$$

Nuclear fission also produces neutrons: most heavy nuclides are fissile, and the fission may be either spontaneous (Section 2.6) or induced. In the latter, a nucleus must first capture a neutron before it splits up; the reaction is abbreviated to (n, f). Since neutrons are released on fission, it is evident that under the right conditions a fission reaction can be self-sustaining. The minimum mass of a fissile isotope needed for this is the *critical mass*: such isotopes must always be handled and stored in sub-critical amounts, to avoid a criticality incident (Section 6.1.2). In a nuclear reactor, the fission of a critical quantity of isotope is sustained in a controlled manner (unlike the atomic bomb, where it is not), and produces an intense neutron flux, along with much thermal energy.

Both of these methods yield *fast neutrons*, whose energies are from about 10 keV upwards. Intermediate neutrons have lower energies, i.e. slower speed; *thermal* neutrons have even less and are approximately in thermal equilibrium with their surroundings. They are produced by *moderation*, or slowing-down, of fast neutrons.

Neutrons have no charge and their interaction with matter is in consequence very different from that of charged particles. It is not electro-dynamic but takes place by means of collisions with atomic nuclei and, since atoms are mostly empty space, this means that neutrons are a penetrating radiation, able to travel for long distances. When they collide with a nucleus, various reactions are possible, e.g. (n, p) or (n, γ), in which the nucleus absorbs the neutron, emits a proton or a gamma ray, and is left radioactive. Through the former reaction, highly ionising particles are produced deep within the irradiated material: if this happens to be a living organism, severe damage can result. Fast neutrons are therefore reckoned to be the most dangerous of all ionising radiations, especially as protection against them is not easy. In sharp contradistinction to charged particles, for which stopping power is proportional to the number of electrons per unit volume, neutrons are stopped in proportion to the number of nuclei per unit volume. Therefore, materials of low atomic number are used for shielding, such as solid hydrocarbons or water; weight for weight, water is some 30 times more effective than lead. Solid hydrogen would be an excellent material but its use is somewhat impracticable!

Many artificial radio-isotopes are made by irradiation of stable elements in the flux of neutrons from a nuclear reactor, moderated to thermal energies since thermal neutrons are the most likely to be captured. Usually the starting and the product nuclei are of the same element, e.g. phosphorus-32 is formed from phosphorus-31 by an (n, γ) reaction; a few light nuclei transmute, e.g. ^{14}N(n, p)^{14}C; ^{35}Cl(n, p)^{35}S; ^{6}Li(n, α)^{3}T. The ability of elements to capture neutrons varies widely, and the *neutron capture cross-section* expresses the probability that a nucleus will capture a neutron: the units are *barns* (so-called from the colloquial expression concerning hitting the broad side of a barn), one barn (10^{-24} cm^2) being approximately the area of an atomic nucleus, whose radius is 10^{-13}–10^{-12} cm. The higher the neutron cross-section of a nucleus, the easier it is to make another isotope from it by neutron irradiation.

2.5 ELECTROMAGNETIC IONISING RADIATION

2.5.1 Sources of the radiation

Electromagnetic ionising radiation originates from several sources, but the physical nature of the radiation from each is identical and in fact the wavelengths (energies) overlap considerably, so that a description of the properties of one applies equally to all.

Cosmic radiation originates in outer space; it is made up of heavy ions which interact with the atmosphere to produce the electromagnetic part of the 'cosmic rays' received on the Earth's surface.

Gamma rays originate inside excited atomic nuclei, often of radio-isotopes: neither mass number nor atomic number are altered by their emission. Their energies are in the range 10 keV up to about 3 MeV (a few have energies up to 7 MeV). *X-Rays* arise from reactions outside the nucleus: they are softer than gamma, having energies up to about 100 keV only. Both are *mono-energetic*, i.e. a given atom always emits radiation with clearly defined energy.

Artificially made radiation results from the bombardment of a target with high-speed electrons accelerated in an electric field in a discharge tube. Radiations produced thus are also called *x-rays*. which seems a pity in view of the possible confusion with x-rays of isotopic origin and the care taken to distinguish these from gamma radiation, especially since Röntgen's rays have historical priority to the name, he himself having called them x-rays. However, the context should make it clear what kind of x-rays are being discussed.

Electrically-generated x-rays are a heterogeneous beam of photons, whose energy spreads from a very low value up to a maximum defined by the high voltage applied to the electrons in the tube: thus a voltage of 100 kV generates photons with energies up to 100 keV, and the radiation is described as 100 kVp radiation, the p standing for the peak energy of the photons. The high voltage is derived by rectification of an a.c. supply, so that there is a certain amount of 'ripple' associated with it and, especially if the rectification is half-wave, this can introduce serious error in accurate work involving very short exposures of the order of milliseconds, such as in diagnostic radiology. A steady voltage is here desirable, and may be obtained by smoothing after rectification, or by full-wave rectification of a three-phase a.c. supply. The radiation is then distinguished by a C following the voltage figure, e.g. 100 kVpC, denoting constant high voltage. Since x-ray machines have a very high power consumption, it is advisable for them to be given their own independent mains supply, to avoid voltage fluctuations.

The softest component of an x-ray beam is usually filtered out, and the *half-value thickness* or half-value layer (HVT, HVL) may also be stated; this is the thickness of a material that reduces the intensity of the beam to half of its original value. Filtration and HVL are customarily stated in terms of aluminium or copper, although water would be of more direct biological application, and would have the advantage that the absorption of x-rays is more nearly exponential in water than in metals, because there is much less hardening of the beam from the Compton effect (*see* the next Section). A final parameter used to define x-ray beam quality is the beam current, expressed in milliamperes (mA).

2.5.2 Interaction with matter

Electromagnetic radiation is uncharged and has no rest mass, so that it cannot ionise atoms by direct interaction with their force fields. However, electromagnetic radiation can interact with matter in at least six ways, of which three produce secondary electrons that are able to bring about ionisations (Figure 2.8), and it is these electrons that give electromagnetic radiation ionising ability.

Photo-electric absorption

A photon is completely absorbed in an electronic shell of the atom and ceases to exist; all its energy is transferred to an orbital electron which is ejected. This interaction is of greatest significance with photons of less than about 0·5 MeV and with nuclei of high atomic number. It cannot occur if the energy of the photon is less than the amount holding the electron in orbit.

Compton scattering

The photon collides with a loosely bound orbital electron, which is ejected as a so-called *recoil electron*. The photon is deflected with its energy reduced by the amount imparted to the electron. This amount is variable, so that recoil electrons may have a range of energies even though they are produced by mono-energetic incident photons. Compton scattering takes place mainly with medium-energy radiation, 0·5–1 MeV, and elements of low or medium atomic number.

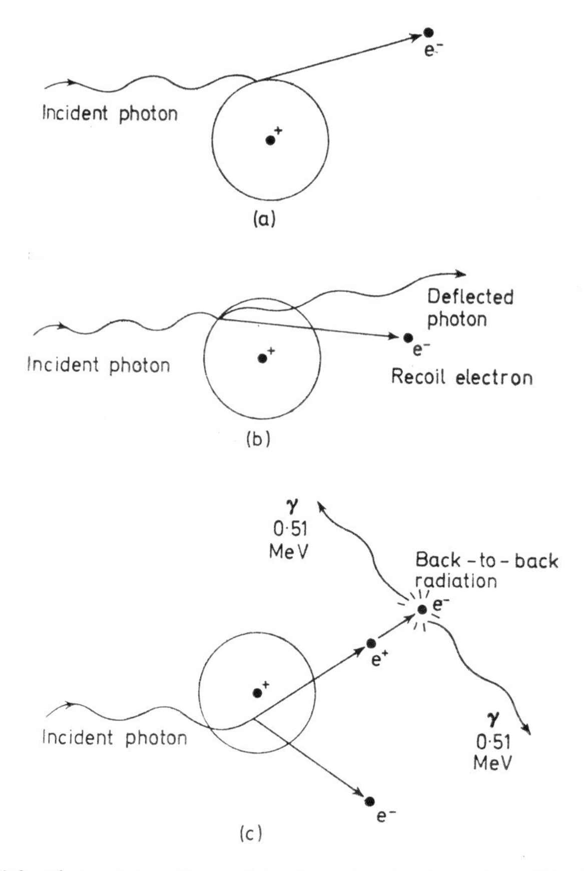

Figure 2.8. Photon interactions: (a) photo-electric absorption, (b) Compton scattering, (c) pair production

Pair production

The photon (here necessarily a gamma ray) reacts with the electric field round the nucleus and ceases to exist, but no ionisation results directly. All the photon's energy is converted, firstly to an electron and a positron, which uses up 1·02 MeV, the equivalent of the rest mass of the two particles, or pair: the surplus energy over and above this is imparted to the pair as kinetic energy. Pair production is more important with low-atomic-number elements and with higher energy gamma photons, whose energy must obviously be at least 1·02 MeV for the process to occur. The positron will give rise to back-to-back radiation as previously described (Section 2.4.2).

Absorption

The absorption of electromagnetic radiation is exponential, so that its range is not clearly defined, and a beam of gamma rays can never be completely stopped or absorbed, but merely attenuated or decreased in intensity. Gamma radiation is much more penetrating than alpha and beta particles: it has a very low specific ionisation, about 1·5/cm in air, since its interaction is indirect.

The mathematical expression for the absorption of gamma radiation by matter is precisely similar to that for radioactive decay, if the thickness of the absorbing material is substituted for time. As the radioactive decay constant is characteristic for a given isotope, so the absorption coefficient is characteristic for gamma radiation of a given energy. Expressed in terms of distance, it is the linear absorption coefficient, μ_1, the fractional decrease in intensity of the beam per unit thickness of absorber: it has the units $(\text{length})^{-1}$.

As with the absorption of alpha and beta particles, the density of the absorbing material is important. This is taken into account in the mass absorption coefficient, μ_m, obtained by dividing μ_1 by the density of the material: μ_m is usually given in units of cm^2/g. Whereas values of μ_1 for different materials are widely different for radiation of the same energy, those of μ_m are very similar. In addition, values of μ_m are independent of the chemical or physical state of the material.

The *half-thickness* of an absorber is the thickness needed to reduce the radiation intensity by half: in the same way as the half-life of an isotope is obtained by dividing 0·693 by the decay constant, so the half-thickness is equal to 0·693 divided by the absorption coefficient. The half-thickness may be defined as either linear or mass, according to which coefficient is used.

When the absorbing material is extensive in all directions, scattering of the radiation is greater, and some may be scattered back into the primary radiation beam. The radiation intensity at a point may then be greater than that indicated by the above simple theory, and the increase is expressed by the *build-up factor* B: B becomes greater as the thickness of material increases. When conditions are such that build-up occurs, they are described as *broad beam*, as opposed to the *narrow-beam* conditions in which B is unity.

If x is the thickness of material,
μ is the absorption coefficient,
I_0 is the initial radiation intensity on the material,

I is the intensity after passing through the material,
and B is the build-up factor,

then,

Under narrow-beam conditions	Under broad-beam conditions
$\frac{I}{I_0} = \exp(-\mu x)$	$\frac{I}{I_0} = B \exp(-\mu x)$
or, $\log_e I = \log_e I_0 - \mu x$	$\log_e I = \log_e I_0 + \log_e B - \mu x$

2.6 DECAY SCHEMES

Radioactive decay results in: first, adjustment of the $N:P$ ratio in the desired direction; secondly, the transmutation of one element into another; and thirdly, loss of energy from the nucleus, so that the daughter is in a lower energy state than the radioactive parent.

The vast majority of isotopes decay by one of four basic processes or modes: three of these are emission of a charged particle, alpha particle, negatron, or positron, and the fourth is *electron capture* (EC), in which the nucleus captures an orbital electron, usually from the innermost or K-shell, giving the alternative name of the process, K-capture. Electron capture is not directly observable, and is manifested as a result of the atom as a whole (in contradistinction to the nucleus) being left with excess energy above its ground state. This energy is released as an x-ray of characteristic energy when one of the other orbital electrons falls into the vacancy left by the captured one, or by the expulsion of one of the outermost electrons as an *Auger electron* (named after its discoverer). A few heavy nuclei (e.g. ^{235}U) decay by *spontaneous fission*, in which the nucleus breaks approximately in half, releasing a small number of neutrons as it does so. The mass of the products is less than that of the original nucleus, and that which is lost is converted into a large amount of energy.

Decay schemes describe how an isotope decays. The simplest are found when the parent goes immediately by the basic decay mode to a daughter nucleus which is in its most stable state, or ground state: this happens with the so-called pure particle-emitting or electron-capture nuclides, such as sulphur-35 (pure negatron), nitrogen-13 (pure positron) and caesium-131 (pure electron capture). Of the few pure alpha emitters that exist, most give daughter nuclides that are themselves radioactive and emit other radiations, e.g. radon-222. Polonium-208 and -210 are almost pure alpha emitters, for they emit a gamma ray in only a very small percentage of their disintegrations.

Simple decay schemes are less common than more complex ones, for the product nucleus remaining after the basic process has occurred often retains some energy above the ground state. Sometimes an internal transition (IT) of the nucleus then releases the excess energy purely as gamma radiation, without any further emission of particles. An alternative to gamma emission is *internal conversion* (IC): the energy of the nucleus is transferred to an inner orbital electron, which is then emitted from the atom with the energy which the gamma photon would have had less the binding energy needed to remove the electron from the atom. These electrons have a discrete energy, and are called conversion electrons to distinguish them from beta particles. X-rays or *Auger electrons* may result from the filling of the resulting vacancy by the remaining electrons.

Usually the basic decay process and the succeeding gamma emission take place practically simultaneously in cascade (e.g. sodium-24), but in a few cases the intermediate excited state can exist for a measurable time; such metastable nuclides are distinguished by an 'm' following the mass number (e.g. technetium-99m). They have a characteristic half-life, like any other radio-isotope. Decay chains are formed when the parent decays to a daughter which is itself radioactive; the longest chains are the natural decay series, and shorter ones are found lower down the periodic table, as for example where a metastable isotope is involved, or as in the decay of strontium-90 to yttrium-90, which in turn yields stable zirconium-90.

Decay schemes are conveniently represented diagrammatically. The diagrams are based on the energy levels of the parent and daughter nuclei, which are drawn as heavy horizontal lines, with a note of the nuclides and their half-lifes. Intermediate energy states (if any) are drawn as thin lines, with a note of their energy above the ground state. The several energy states are linked by thin lines representing the nuclear changes. They are drawn diagonally down to the left if the daughter nucleus is lower down the periodic table (electron capture, positron emission, alpha emission), to the right if it is higher up (negatron emission), and vertically down if there is no change in position (gamma emission). The energy of each radiation is indicated. If there is more than one possible route for decay, the percentage of disintegrations following each is also shown, since it is found that these percentages are constant. Decay schemes can be very complex, especially where more than one daughter element is formed (e.g. bromine-80 which decays

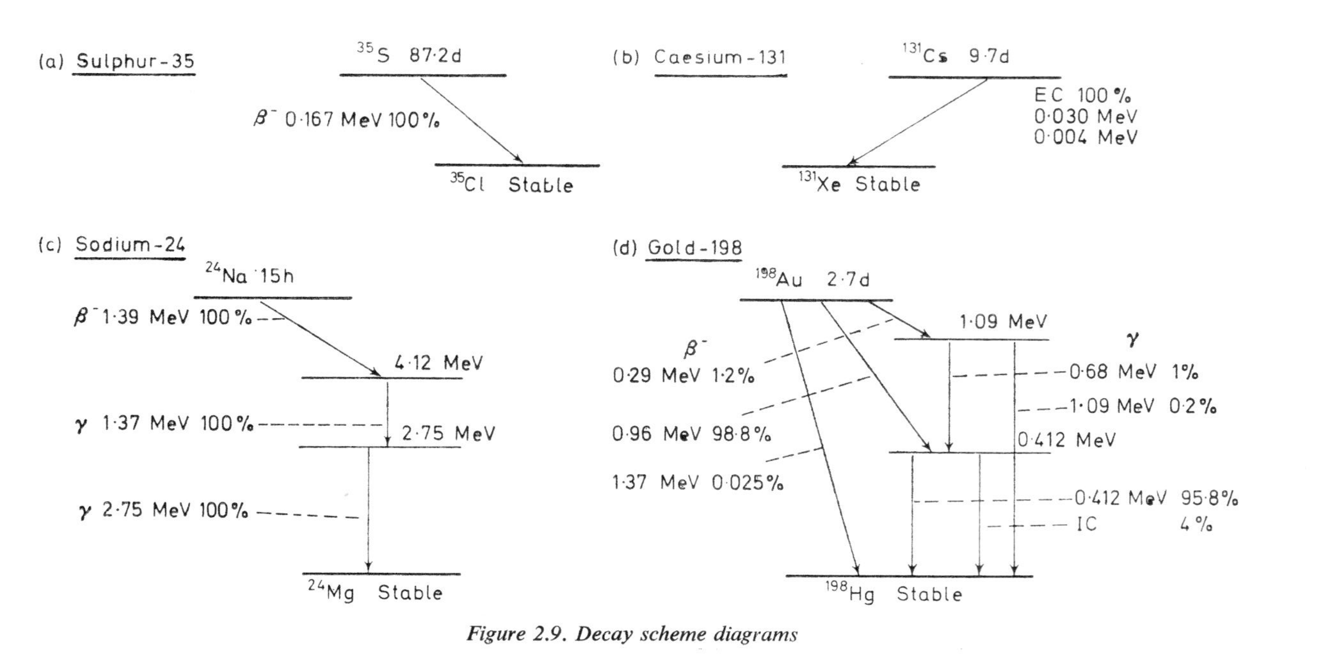

Figure 2.9. Decay scheme diagrams

by negatron emission or positron emission or electron capture, and has a metastable state to complicate matters), or where a large number of alternative routes are possible, e.g. tantalum-182. Usually it is possible to draw only the main routes in such cases. Figure 2.9 gives examples of some decay scheme diagrams.

The Group Displacement Law is useful in working out decay schemes: for example, it is known that loss of an alpha particle will always result in a product nucleus two places lower down the periodic table (since two positive charges are carried off by the alpha) and four mass units lighter. Similarly, emission of a negatron always moves the nucleus one place higher in the table with no change in mass. The Law has now been extended to include all nuclear reactions, bombardments as well as decays (Figure 2.10).

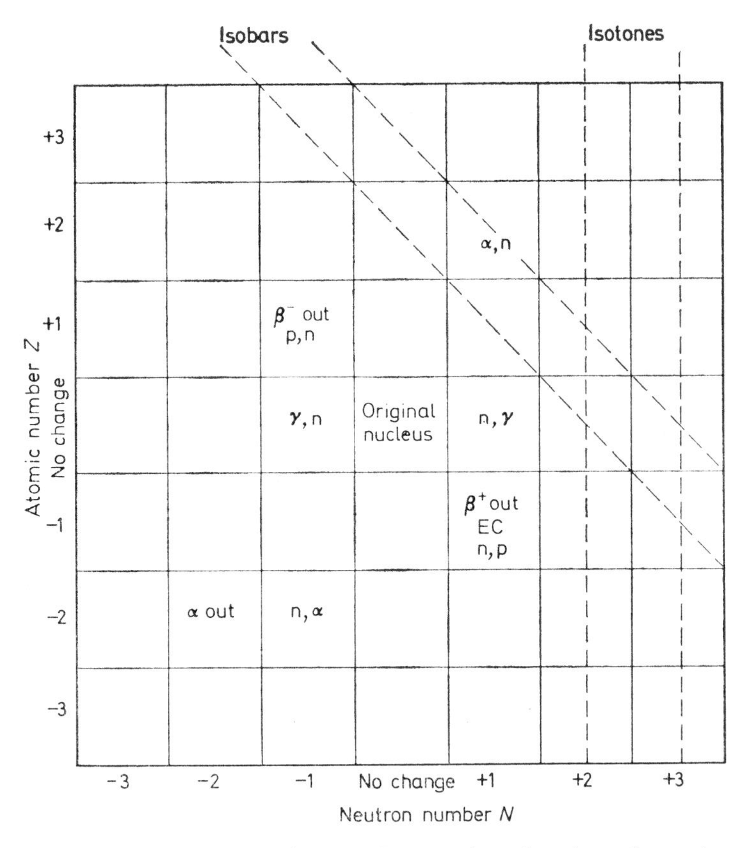

Figure 2.10. The Group Displacement Law: nuclear alterations after various changes

Chapter 3

THE DETECTION AND MEASUREMENT OF IONISING RADIATION

3.1 GENERAL CONSIDERATIONS

The three most common methods for detecting and measuring ionising radiation are based on the production by the radiation of first, ionisation in gases; secondly, excitation in solids or liquids; and thirdly, chemical change. A fourth method, the production of ionisation in solids, has been developed only comparatively recently, and is mentioned at the end of the chapter. It should perhaps be made clear that not all solids, liquids, and gases can be used for detecting radiation—only a few are suitable in practice.

Radiation may be measured in either the *differential* or the *integral* mode. The former records the number of individual ionising events, differentiating (or distinguishing) between one event and another, whereas the latter measures their total cumulative effect, integrating (or combining) the individual effects from the separate events.

The measurement of radiation in radiotracer work, commonly called *counting*, is concerned with the determination of activity in terms of disintegration rate, and is most often a differential process. By contrast, the measurement of radiation in health physics, *dosimetry* (Section 6.3), is in terms of the dissipation and absorption of energy in biological systems, and is almost always an integrating process.

Absolute counting, the detection of every disintegration event in a radioactive sample, i.e. measurement of its disintegration rate, is very difficult. Counting operations are therefore almost always relative, detecting only a proportion of the disintegrations in a sample and measuring its count rate. The detector may be calibrated with a source whose activity is precisely known, if it is desired to know the activity of a sample very accurately, but in many experiments all that is required is to relate the count rates of the unknown to a standard preparation, e.g. an aliquot of the isotope stock solution that was put into the experiment at the beginning, and there is then no need for such a calibration.

However, any comparisons will only be valid so long as all counts are taken under identical conditions, i.e. *counting geometry* must be constant. Geometry refers to the arrangement of the sample with respect to the detector, and can be taken to include the chemical and physical form of the sample, and the way in which the sample is supported and shielded. Two common terms are '2π' and '4π' geometry; they are derived as follows. A detector able to pick up radiation emitted in any direction from a source may be regarded as doing so over the whole surface of a sphere having the source at its centre. The surface of a sphere of unit radius is 4π square units, and such a detector is therefore described as having 4π geometry. Likewise, a detector which picks up radiation over the surface of a hemisphere above the source has 2π geometry.

3.2 GAS IONISATION COUNTING METHODS

3.2.1 Principle

A gas ionisation detector, or ion chamber, is basically an earthed chamber which contains, insulated from it, an electrode at a positive potential, i.e. an anode. All gas ionisation detectors are based on this principle, but their detailed designs differ according to the quality and quantity of radiation and the conditions of use for which they are intended, and in general it is not normally possible to use an ionisation chamber detector satisfactorily, if at all, unless these conditions are adhered to. Ionisation chambers are best suited to the detection of particles whose specific ionisation (Section 2.4.3) is high, i.e. alpha and beta radiation. The specific ionization of gamma radiation is low, and a gamma photon may pass right through an ionisation chamber without causing any ionisation at all, and consequently the efficiency of detection of gamma radiation by gas ionisation methods is low.

3.2.2 Simple ionisation chambers

The simplest gas ionisation detector is the *electroscope*, which is very similar to the traditional gold-leaf electroscope familiar in elementary electrostatics, and in fact it works on the same principle. One end of a light fibre of metallised quartz is attached to the anode; mutual repulsion causes the fibre to move away from the anode, and the amount of charge on it is measured in terms of the angular displacement of the fibre. The gas in the electroscope ionises as the radiation passes through it, and the ions migrate to the electrode of opposite sign, where they are discharged. The discharge of the electrons reduces the charge on the anode, so that the fibre moves back towards it; the movement gives a measure of the amount of discharge and therefore of the amount of ionisation that has taken place. The electroscope is an integrating instrument.

The simple electroscope takes rather a long time to record low activities, so that it is not of great use for experimental work today (though be it remembered that the early pioneers in radioactivity had little else, and carried out all their experiments with it). At the same time, however, it is useful in that it can measure high activities without becoming swamped, unlike other types of detector, and it can be made in small, portable sizes, so that it is useful in dosimetry (Section 6.2). A larger type is the *Lauritsen electroscope.*

If the anode is maintained at a constant potential by an external circuit, the discharge of the electrons it collects generates a flow of current, the ionisation or *ion current*, whose magnitude is directly related to the number of ionising particles being detected. It is not a steady current, but is made up of a succession of pulses, since whereas the heavy positive ions move relatively slowly towards the cathode, the light electrons move very quickly to the anode, and those produced by one ionising particle arrive essentially all at once regardless of where in the chamber they were formed. Their simultaneous discharge then produces a momentary surge or pulse of electricity. In differential counting, the external circuit counts the number of these pulses; integral counting measures the magnitude of the ion current.

The size of the pulse produced by a single ionising particle may be easily calculated, as follows. About 30 eV are needed for an alpha to form an ion pair in a gas, so that a particle of high energy, for example an alpha of 6·0 MeV, will produce about 2×10^5 ion pairs. One electron carries $1{\cdot}6 \times 10^{-19}$ C, so that the arrival of this seemingly large number of electrons at the anode will generate a charge of only $3{\cdot}2 \times 10^{-14}$ C. It is therefore apparent that ion currents, even with large fluxes of energetic particles, are very small, and will be even more so with the smaller fluxes of lower energy particles that are more commonly encountered, as with the beta-emitting isotopes used in tracer experiments. Such small currents are difficult to measure directly, and in instruments other than the electroscope they are amplified before measurement in order to make them easier to handle, either within the chamber (see below) or outside it, as in the electrometer.

However, another problem at once arises, for the ion current is direct current, and d.c. amplifiers are inherently unstable with small currents. Electrometers which amplify the ion current directly, therefore, do not have a high degree of precision. The *vibrating reed electrometer* (VRE) overcomes the problem; the ion current is collected on a capacitor, one plate of which is rapidly vibrated mechanically at a frequency of several hundred cycles per second (Hz), so that its capacity is constantly varied and its output becomes converted to the more easily handled a.c. After amplification, the a.c. is rectified back to d.c. for measurement. The VRE has several advantages: first, it is versatile, for it can deal with solid, liquid or gaseous samples without modification, and its efficiency of detection approaches 100% with gaseous samples. Secondly, it has a wider working range than any other detector, for it can be used to measure activities from some 50 pCi to about 1 mCi. Thirdly, it can be made portable, and it is therefore useful in dosimetry.

3.2.3 Ionisation chambers with gas amplification

It was mentioned earlier that there is another way of amplifying the ion current other than by an external amplifier circuit; this is by means of gas amplification inside the ionisation chamber itself, which may be understood by considering the characteristics of ionisation chambers, i.e. how the charge collected at the anode varies with the applied voltage.

When the voltage is zero, the charge collected is likewise zero, because there is of course no force to draw the electrons to the anode, and they will all recombine with the positive ions. If a potential gradient is applied, the electrons will move towards the anode, and some will reach it before recombination with positive ions can take place. As the voltage increases, the electrons will move faster, and more will reach the anode, so that the charge collected increases with the voltage up to a point at which all the electrons are being collected before they can recombine. The charge received at the anode then levels out to a constant value, the *saturation current*, which is virtually independent of the applied voltage. A simple ionisation chamber operating in this region does not therefore need a highly stable power supply. The saturation current plateau for alpha particles is higher than for betas (Figure 3.1), since they are of higher initial energy, so that it is possible to

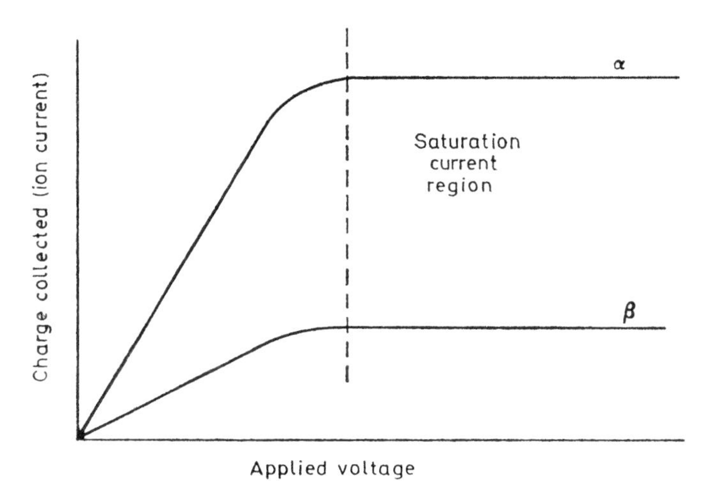

Figure 3.1. Characteristics of a simple ionisation chamber

distinguish the two, according to the size of pulse they produce, by a simple form of *pulse height analysis* (PHA).

If the voltage applied to the ionisation chamber is now further increased, the ion current no longer remains constant, but again starts to increase, and the saturation current region is left behind (Figure 3.2). This is because the higher voltage accelerates the electrons to such a speed that they themselves can act as ionising particles (delta-rays) (Section 2.4.3) and produce secondary ion pairs, which in turn can bring about still more ionisation. The

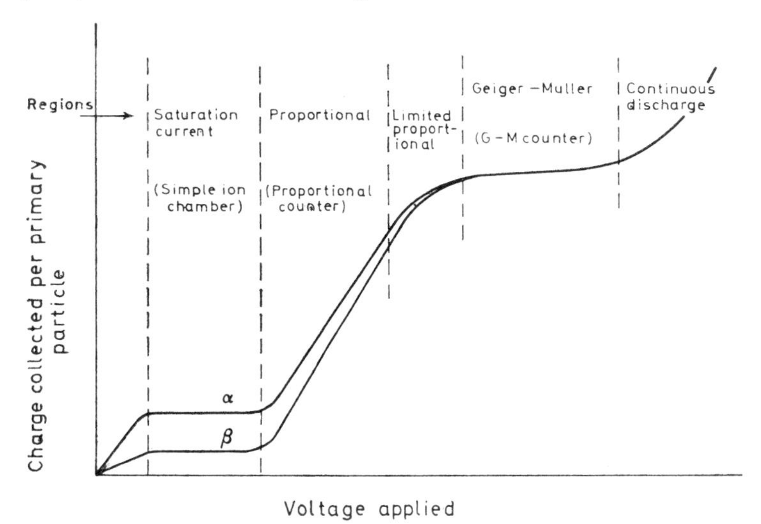

Figure 3.2. Characteristics of an ionisation chamber with gas amplification

number of ionisations thus increases in geometrical progression, so that the number of electrons arriving at the anode is likewise increased, and the ion current is thus greatly amplified within the ionisation chamber. This is the phenomenon of gas amplification. The near-simultaneous arrival of this large number of electrons at the anode is named the *Townsend avalanche*, after its discoverer, a most apt description that could hardly be bettered. Some further degree of amplification of the pulse externally may still be needed, but it is much less than with the simple ionisation chamber, and is also much easier since the current to be amplified is itself larger. The gas amplification factor (A) expresses the ratio of the charge finally collected at the anode to that produced by the primary ionising event, or that which would have been collected without the gas amplification. In other words, A is the number of ionisations finally produced from one primary ionisation. Its value depends on the gas filling the counter and its pressure. Argon is one of the most readily ionised gases, and it and the other 'inert' gases are often used in ionisation chambers with gas amplification. Methane may also be employed. It is important that the filling gas should be free from oxygen or chlorine, since electrons attach to these gases, forming heavy negative ions; these would obviously affect the performance of the counter adversely, and are undesirable.

Gas amplification introduces the concept of *dead time*. Although the electrons of the ion pairs are effectively discharged all at once, the heavy positive ions are not, and remain as a sheath round the anode, which destroys the electric field. Consequently the counter is completely insensitive and unresponsive to the entry of a second ionising particle until the positive ions begin to move outwards to the cathode and become discharged. This is the dead time of the counter. During the *recovery time* the electric field begins to re-assert itself as the ions discharge, and the sensitivity of the counter gradually returns to normal. As it does so, it will respond to a second particle with a pulse whose size increases likewise; how soon this second pulse is picked up by the counting circuit would depend on the setting of the *bias*, (Section 3.5). The time following the detection of one ionising particle before another can be recorded is the *resolving time*, *paralysis time*, or *quenching time;* it is also loosely known as dead time. Two particles entering the counter within this time of each other will not be detected as separate events but will be recorded only as one. Thus the existence of the resolving time places an upper limit on the intensity of radiation that the counter can detect, and means that corrections for *lost counts* have to be applied at high count rates, or where the dead time is long. This is notably so with Geiger–Muller counters. Lost counts are not generally significant where dead time is short, as in scintillation counting.

Four regions may be distinguished above the saturation current part of the ionisation chamber characteristic curve. They are the proportional, limited-proportional, Geiger–Muller and continuous discharge regions.

The proportional region

At a fixed setting of the high voltage applied to the chamber, the gas amplification factor A is constant. The number of ion pairs finally produced and the charge collected at the anode is directly proportional to the energy of the particle entering the chamber, and it is thus possible to differentiate between alpha and beta particles, and even between those of the same kind if their energies are sufficiently different. If an alpha forming 10^5 primary ion pairs enters a chamber whose A is 10^3, 10^8 electrons will be collected at the anode, whereas a beta able to give 10^2 primary ion pairs initially will generate only 10^5 electrons at the anode.

Counters working in this region are termed proportional counters. Their voltage supply has to be highly stable for accurate work, but they have some compensating advantages: they can easily be used for radioactive gases, they can be used for the detection of very low energy particles, and their resolving time is short (5–50 μsec) so that very high counting rates are possible. Their A values are about 10^2 to 10^4.

The limited-proportional region

Here, the transition is made from the proportional to the Geiger–Muller region. The value of A increases with the voltage up to a point beyond which strict proportionality is not maintained between the energy of the particle and the charge collected. This is because the amplification has an upper limit, imposed mainly by the finite number of ionisable gas atoms in the counter. Once they are all ionised, it is obvious that no more ions can be formed. If the maximum possible number of ions is 10^{10}, and A is 10^7 (for a single ionising event) the same beta particle as above will be amplified to give 10^9 electrons; the alpha, however, cannot give the 10^{12} that it should, but is restricted to the available maximum of 10^{10}. The distinction between different energy particles is thus greatly lessened. Counters are not normally operated in this region.

*The Geiger–Muller region**

The value of A is now so large that even the weakest ionising particle, and in fact a single ionising event, will produce the maximum possible number of ions. The size of the pulse is now completely independent of particle energy, and is governed by the physical properties of the counter itself. It is also independent of the applied voltage over a fairly wide range, until the final region of *continuous discharge* is reached, in which ionisation, once started, will not stop. The A value for a Geiger–Muller counter is about 10^8.

3.2.4 The Geiger counter

There can be few people who have not heard of a Geiger counter, or who have no idea that it has something to do with radiation, for the Geiger–Muller (G–M) tube is the most commonly used form of detector. This is not without good reason, for it can be made simple and robust in construction, and fairly cheaply; its high voltage supply does not have to be highly stable, and the voltage output is large, within the range 0·1 V to (for some halogen-quenched counters) 50 V, so that little or no amplification externally is required before counting. Even the weakest energy beta particles are detected with 100% efficiency, provided that they can get into the sensitive volume of the counter (see below). By contrast, the efficiency of detection of gamma radiation is only about 1%, but until the development of scintillation counting there was little alternative to making the best of a bad job and using G–M tubes designed to be as efficient as possible for gamma radiation. It may be noted in passing that ultra-violet light can excite such tubes, e.g. exposure to bright sunlight without being screened, and this can be misleading if it is not noticed.

There are snags, however, the first being that, in comparison with other detectors, the resolving time is long, 100–200 μsec, because the larger number of positive ions takes longer to reach the cathode wall and be discharged. The resolving time is also slightly variable, so that the necessary corrections for lost counts in accurate work become complicated. G–M tubes are therefore usually connected to a *quenching pre-amplifier*, sometimes called a *probe* unit, which introduces a constant, though longer, dead time.

* Muller should strictly speaking be spelt with an umlaut over the u (ü) though this is often omitted.

The second snag also arises from discharge of the positive ions, not so much this time because there are more of them, but because the neutralising electron from the cathode tends to be in a higher energy state than the one removed in the ionisation. As this electron falls to the ground state, the excess energy is released as a soft x-ray and, in spite of the low efficiency of the G–M tube for this sort of radiation, the x-ray would be enough to promote a new ionisation avalanche, so that the tube would go into a state of perpetual ionisation if some means of *quenching* it were not used. There are two possibilities: first, an external quench, in which the potential across the tube is reduced momentarily after each pulse until all the positive ions are safely neutralised. This accentuates the dead time. A more recent development is the internal quench. A gas with a lower ionisation potential than the inert filling gas is introduced into the tube. The latter is ionised by the radiation as usual, but its ions are now discharged at the expense of the quench gas, which becomes ionised instead—apparently the radiation does not ionise the quench gas directly. The quench gas ions are then discharged at the cathode wall, where their surplus excitation energy is expended in chemical degradation or dissociation of the molecule, and is not released as electromagnetic radiation. The discharge of the tube is therefore not perpetuated.

A suitable quench gas can be either an organic compound or a halogen. The former comprise about 10% by volume of the filling of the G–M tube, and typical examples are ethyl alcohol or ethyl formate. Their breakdown is irreversible, so that they are used up gradually, and the tube has a definite life: for this reason, an organic-quenched tube should always have sources removed or the high voltage turned off except when a count is actually being taken, so as to get maximum use from it. An average-sized tube will contain about 10^{20} molecules of quenching gas, of which about 10^{10} are used up each pulse, so there are theoretically sufficient for 10^{10} pulses, though in practice the diminishing concentration of quenching gas determines the life at about 10^7–10^9 counts.

Halogen-quenched tubes normally use bromine. Here, the molecules simply dissociate and the products can readily recombine, so that the gas is not used up and the life of the tube is indefinite. They are rather more difficult to manufacture, for very little bromine vapour is required and, as it is very reactive chemically, it tends to disappear into the electrodes and sealing materials. Their background counting rate (Section 4.5.3) is slightly higher than for organic tubes.

It is necessary to know the characteristics of a G–M counter to use it correctly, i.e. the variation of the count rate received from a source with the applied potential (Figure 3.3). At first, no counts whatever are recorded, the counter behaving as a simple ion chamber. As the voltage is increased, gas amplification starts and the counter suddenly begins to work as the starting and threshold potentials are reached. The slope of the graph is here

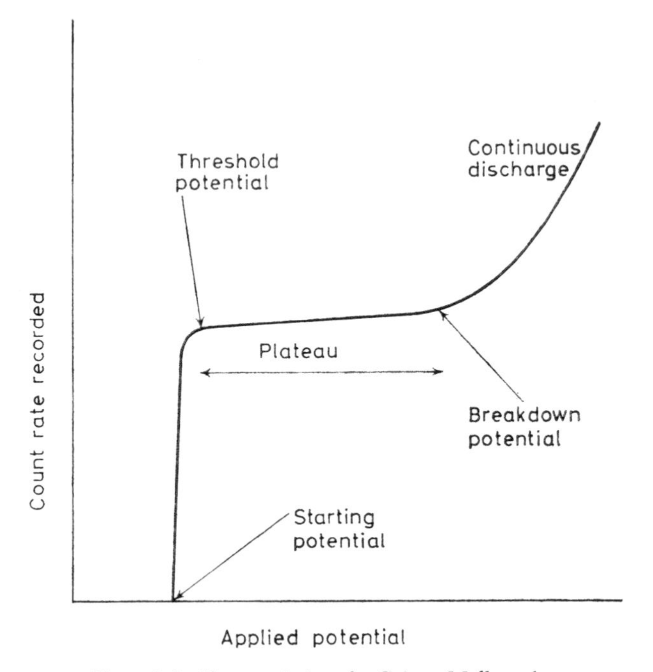

Figure 3.3. Characteristics of a Geiger–Muller tube

very steep. It then flattens out on to the *plateau* as far as the *breakdown potential*, after which the tube goes into continuous discharge: it should never be allowed to remain in this region, or it may be severely damaged. The counter is operated in the plateau region, and since the count rate here varies very little with the applied potential, the high voltage supply to a G–M tube need not be very highly stabilised.

Types of Geiger counter

There are many different designs of G–M tube, and only a few that are of general interest and application can be described. The commonest is the *end-window* or '*bell*' type, Figure 3.4, a cylinder with the anode entering at one end and closed at the other by a 'window' of thin aluminium or split mica. The window thickness is always expressed in mg/cm²: for metal, it is about 7 mg/cm² and for the thinnest mica, 1–2 mg/cm². The filling gas is at a pressure of some 100 torr, which causes the thin window to bow inwards. The cathode connection is via the wall of the cylinder;

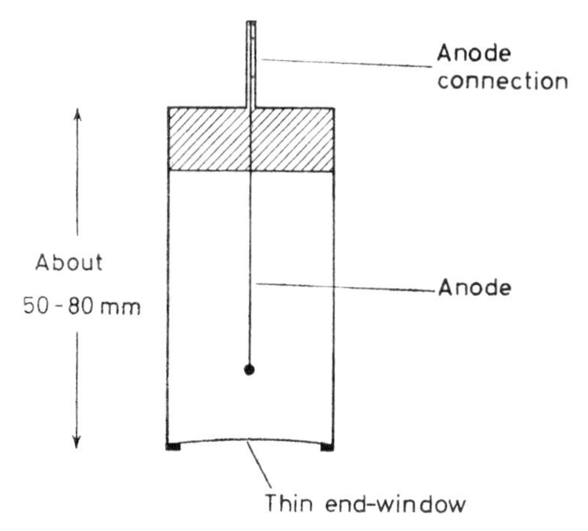

Figure 3.4. End-window Geiger counter

alternatively, both anode and cathode connections can be via a valve base. Samples are counted on a small shallow aluminium dish, or *planchet*, placed below the window. (The planchette used in spiritualism is quite different.) The existence of a window interposes a barrier between the sample and the sensitive volume inside the tube, and thereby creates a difficulty in the detection of alpha and low-energy beta particles, which are absorbed significantly or completely in even the thinnest mica windows (Section 2.4.3): several biologically important isotopes (e.g. tritium, carbon-14, sulphur-35, and calcium-45) are pure soft beta emitters falling in this category.

Such soft beta emitters (and alpha emitters) are therefore better assayed in a proportional or a *windowless flow counter* (Figure 3.5), in which the sample is placed within the sensitive volume of the detector, and there is no intervening window. A constant flow of gas is passed through to exclude air, a typical example of such a flow gas being 2% isobutane in argon. Gas enters the counting chamber, passes through a hole in the turntable to the pregassing chamber, then out through the bubbler above, which indicates the flow rate. The O-ring ensures that the anode unit is a gas-tight fit in the counting chamber. The turntable has three positions: entry–exit; pregassing, to sweep out air introduced with the sample;

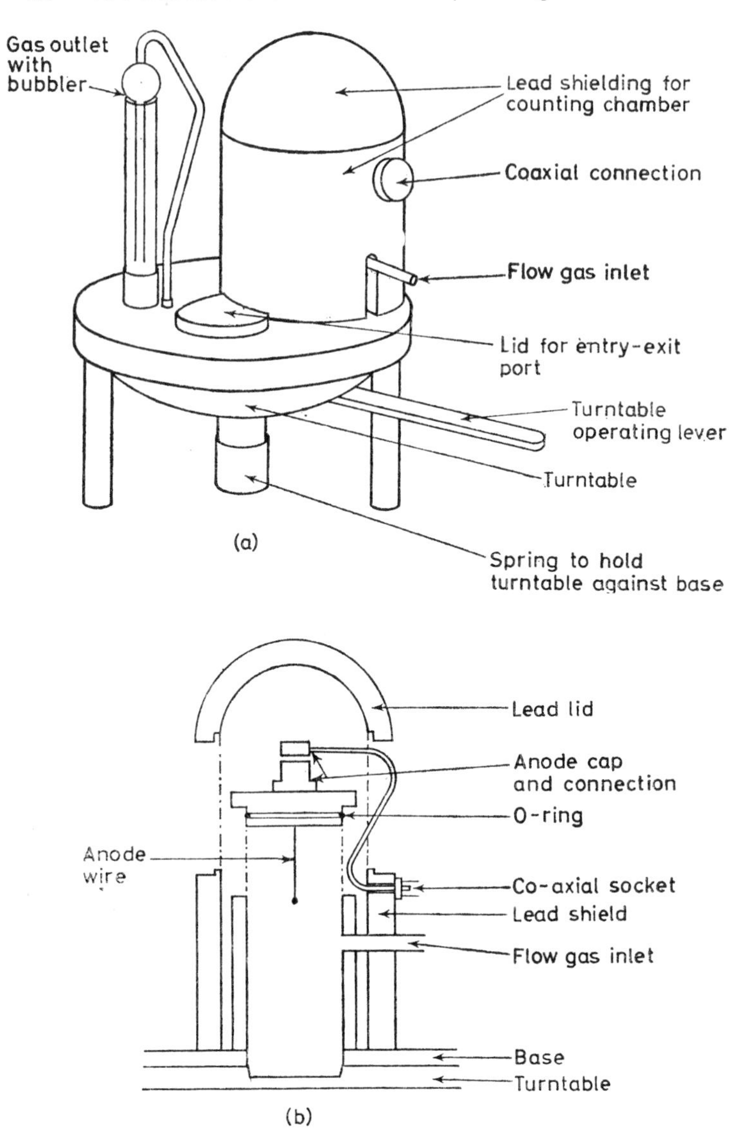

Figure 3.5. Windowless gas-flow Geiger counter—size about 230 mm *high, 150* mm *across: (a) perspective sketch, and (b) cross-section of counting chamber*

and counting. The cathode connection is via the wall of the chamber. Scintillation counting is an alternative means of detection.

Ultra-thin windows of plastic ('Mylar'), 150 μg/cm², which will transmit even alphas, have been developed recently, but detectors using them have to be operated as flow counters, with a constant input of filling gas at about atmospheric pressure to counteract the inward diffusion of air that would otherwise occur.

Another type of tube is made of glass; longer and narrower than the end-window design, it is designed to give the best possible detection of gamma radiation.

'*Liquid*' counters are also made of glass, and are intended to be either dipped into a radioactive liquid, or to be surrounded with liquid—here, the thin glass wall is surrounded by an outer thicker one, leaving an annular space of about 5–10 ml capacity into which the liquid is poured (Figure 3.6). This type may be used with

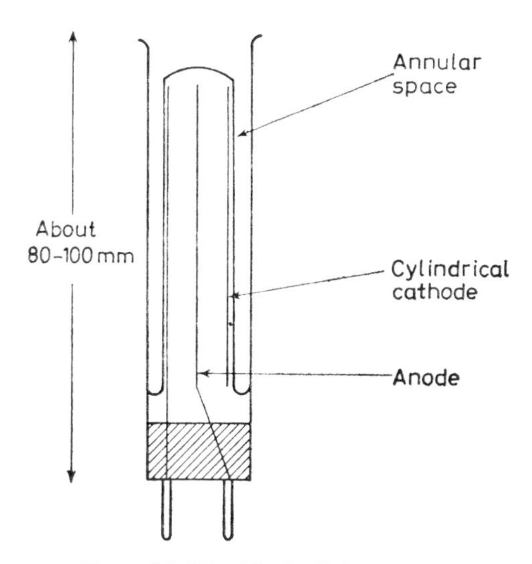

Figure 3.6. 'Liquid' tube Geiger counter

a flow of liquid, in which case the construction is very like a Liebig condenser, with the detector in the middle and the liquid flowing round it.

A type designed for easily-condensed *gases*, e.g. carbon dioxide or water vapour, has an extension at the bottom that can be cooled in liquid nitrogen (Figure 3.7). The gas to be counted is passed in through an inlet at the top and condenses in the cold finger. Filling gas is then admitted and the tap is closed; the tube is attached to the electrodes of a counting system, and when the condensed gas has evaporated and mixed with the filling gas, it is counted. The geometry here is 4π, whereas in the other types of G–M tube mentioned it is 2π.

An *automatic scanner* is frequently used for counting chromatograms of radioactive substances. The chromatogram is moved slowly under an end-window or windowless-flow G–M detector (or vice-versa), whose output is linked to a ratemeter (Section 6.4.1.) and a strip chart recorder.

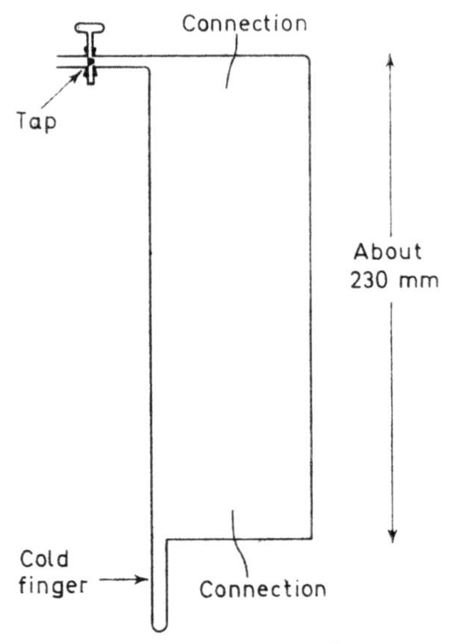

Figure 3.7. Geiger counter for gases

3.3 SCINTILLATION COUNTING

3.3.1 Principle

Scintillation counting utilises the principle that matter may be excited by the passage of ionising radiation through it. An orbital electron is raised to a higher energy level and almost immediately drops back to its ground state, releasing its excitation energy as electromagnetic radiation. The energy of the excited state above the ground state varies from one substance to another, so that the amount of excitation energy released likewise varies, as also will its wavelength, since energy and wavelength of electromagnetic radiation are inversely related. (Section 2.1.2). The energy relations in certain substances are such that the wavelength is in the visible range, and the passage of radiation through them is therefore manifested by the appearance of minute flashes of light, or scintillations, one flash corresponding to the passage of each ionising particle or ray. Such materials are called fluors, or scintillators; they are sometimes also known as phosphors, but this term is best avoided to avoid confusion with the rather different phenomenon of phosphorescence. The secondary electrons produced when gamma rays interact with matter bring about the excitation of the fluor, not the gamma itself.

3.3.2 Detection of the scintillations

Scintillations can be seen (p. 9) but are now detected by a *photomultiplier* (PM) which is basically a series of electrodes, or *dynodes*, each maintained at a positive potential with respect to the preceding one. In front of the dynodes is the light-sensitive cathode, or photocathode; this is a layer, so thin as to be transparent, of light-sensitive material, typically caesium or a caesium–antimony alloy. When light strikes the photocathode, electrons (photo-electrons) are emitted, and they are accelerated towards the first dynode by the applied potential gradient. *Secondary electron multiplication* then takes place: the photo-electrons strike the surface of the dynode with sufficient force to knock out a larger number of electrons. These in turn are accelerated towards the next dynode, where the process is repeated, and so on down the whole series. There is thus an amplification of the original photo-electrons in geometrical progression along the PM, and the minute amount of incident photon energy is converted into a sizeable pulse at the final collecting anode. There may be 11 or more dynodes in the PM, so that a gain of only two or three at each stage can nevertheless result in an overall gain of several thousands. The voltage applied to the PM governs the amplification; it is spread out evenly across all the dynodes, giving 100–200 V for each stage, and can be adjusted as desired. The charge produced at the final anode by a radiation of given energy can easily be calculated if various parameters are known, namely, the efficiency of conversion of radiation into photon energy, the efficiency of the photocathode, and the multiplication at each stage of the PM. Although greatly amplified, the pulse is still too small to be counted directly, and it is then fed into a conventional linear amplifier before counting.

3.3.3 Quantitative considerations — gamma-ray scintillation spectrometry

So long as conditions are constant, a quantitative relationship holds all the way through scintillation counting, first between the energy lost by the incident radiation and the excitation energy imparted to the fluor. This then determines how much photon energy is released when the fluor reverts to its ground state, which in turn governs the size (voltage) of the pulse that emerges from the PM. These pulses form a spectrum, whose shape (expressed as the graph of count rate, i.e. number of pulses in unit time, against pulse height) depends on how the radiation loses its energy to the fluor.

For medium-energy gamma radiation, this energy loss takes place principally by the photo-electric effect and the Compton effect. In the photo-electric effect, the whole of the gamma ray energy is lost to the fluor and, since gamma radiation is mono-energetic, the resulting pulse spectrum is a single peak with a statistical spread, centred on a particular value (the photopeak). In the Compton effect, only part of the gamma energy is transferred to the fluor, and the spectrum forms a continuous band, the '*Compton smear*', extending up to and merging into the photopeak. Pulse heights may be varied by altering the voltage on the PM and/or the gain of the amplifier, according to the needs of the experiment. They are independent of the quantity of isotope present in a sample; increasing the latter merely increases the pulse frequency, i.e. the count rate (Figure 3.8).

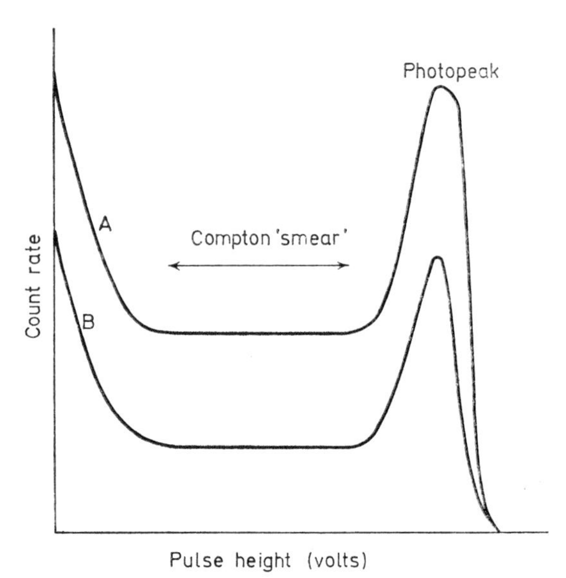

Figure 3.8. Diagrammatic gamma scintillation spectra: A—*large quantity of isotope;* B—*small quantity of isotope*

The spectrum may be complicated by pair production, and interpretation may become difficult if two or more isotopes are present, but nevertheless it is possible to sort out and count separately the photopeaks corresponding to the different energy gamma rays. This is gamma ray scintillation spectrometry, or pulse height analysis, and it is carried out with the aid of a pulse height analyser (or '*kick-sorter*'). The analyser is basically two discriminators; one passes only those pulses greater than its setting, and the other passes only those which are less. Both are variable, and the result is a variable width '*gate*' or '*window*' that can be moved anywhere in the spectrum. Only those pulses falling within the selected gate will be passed on to the counter. If the gate is set so as to straddle the photopeaks of particular isotopes in turn, they can be counted without interference from each other (assuming that the photopeaks do not overlap). It is possible, though less convenient, to carry out pulse height analysis with only one discriminator, counting first with it set to the bottom of the desired gate, then with it set to the top, and subtracting the two counts.

Scintillation spectrometry is a most valuable technique with mono-energetic radiations, particularly gamma (it also has application with alphas), but it is much more restricted with beta-radiation, whose energy is spread over a range. It may be used for quantitative analysis of a mixture of known isotopes, and for identification of an unknown, by comparison of its photopeak with those of isotopes of known energies since, if the spectrometer is working properly, the graph of photopeak height against corresponding photon energy is a straight line through the origin (Figure 3.9). The *resolution* in gamma spectrometry is defined as the full width of the photopeak at half its maximum height (FWHM).

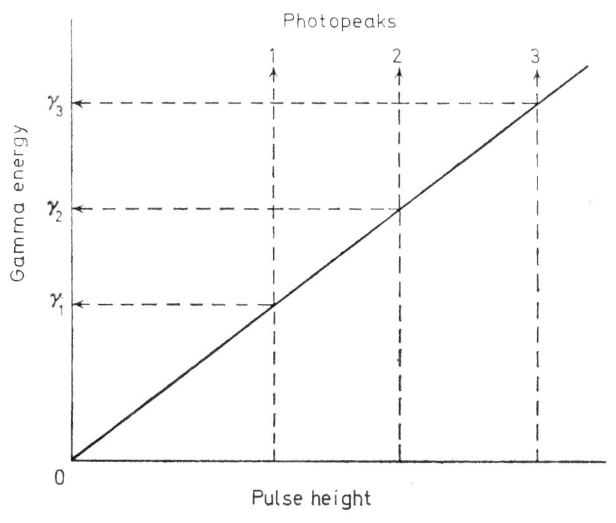

Figure 3.9. Relation between gamma photon energy and photopeak height

Scintillation counting can record very high count rates, for the decay time of the photons is very short, and the time taken for the PM to process one pulse is also very short; consequently, the resolving time is very brief, of the order of a few microseconds. However, *background* is inherently higher than in gas ionisation methods, arising from thermally generated electronic noise in the circuitry and the high amplification in the photomultiplier. It can be greatly reduced or almost eliminated in various ways, mainly by pulse height analysis. *Cooling* the photomultiplier helps greatly; cold water from the tap is often adequate, but for the lowest energy radiations, such as that from tritium, it may be necessary to place the whole counter in a refrigerator or deep-freeze. *Anti-coincidence circuitry* (Section 3.5) may also be used.

A widely-used criterion for assessing the performance of a scintillation counter is the 'E^2/B' figure, i.e. the square of the source count divided by the background. The controls of the counter are usually set so as to make this a maximum (Roberts, 1967).

3.3.4 Methods of scintillation counting — external-sample counting

The relative geometrical arrangement of sample and fluor takes one of two main forms: external- and internal-sample conditions. In the former, as the name suggests, the sample is external to the fluor; in the latter, fluor and sample are in solution together. In all cases the system must be operated in complete darkness, for the PM will be ruined if daylight or room lighting reaches it while the high voltage is applied.

External-sample counting uses a single solid crystal of fluor, optically coupled (e.g. with silicone oil) to the face of a photomultiplier. If the sample is simply placed in contact with one face of the crystal, a maximum of only 50% of its radiation can enter the crystal, the remainder travelling away from it, and the geometry is 2π. A *well crystal* is one that is made with a hole, or well, in which the sample is placed, so that much more radiation enters the crystal and the detection efficiency is correspondingly higher: the geometry is nearly 4π. The fluor crystal may be inorganic or organic, the choice, and the dimensions, depending on the type of radiation being detected. Inorganic crystals contain a small controlled amount of impurity, or '*activator*', that is essential for the functioning of the crystal as a scintillation detector. The energy of the radiation is transferred from the fluor to the activitor, and its de-excitation gives rise to the scintillations.

Fluors for gamma detection

A typical inorganic scintillator for gamma detection is sodium iodide containing about 1% of thallous iodide, used in thick crystals, 30 mm square or larger. The crystal is housed in a thin aluminium container that protects it from external light and mechanical damage, and is lined with powdered magnesium oxide, forming a white surface to reflect the scintillations into the PM. The face of the crystal which fits against the PM is sealed with a glass or Perspex window, for protection against mechanical damage and (more importantly) against moisture, for sodium iodide is hygroscopic and the crystal would rapidly be ruined by contact with air.

Fluors for alpha detection

Zinc sulphide activated with silver is used for alphas. The size of the crystals is important, as they are opaque to their own radiations; the optimum thickness is about 8 mg/cm^2. They are supported on a transparent base placed in close contact with the photocathode of the PM, and are covered with very thin aluminium (3×10^{-5} in, about 0·2 mg/cm^2), which excludes light but allows the alphas to pass, two thicknesses usually being used because of the difficulty of avoiding pinholes in such thin foils.

Fluors for beta detection

Organic scintillators used for beta radiation are anthracene, or naphthalene containing 0·1% anthracene. An alternative to the organic single crystal is to impregnate polymers with the organic material, in which form they are known as *plastic fluors.* Although they are not very good for spectrometry, they can be used for specialised work, as they can be machined into shapes that would be difficult or impossible to make with single crystals. Soft beta radiation is usually determined by internal-sample methods (see next section) but Rosen, Lauree, and Eisenbud (1967) have counted carbon-14 and tritium by means of the bremsstrahlung they produce.

3.3.5 Internal-sample or liquid scintillation counting

Inorganic compounds are able to act as fluors only in the solid state, whereas organic fluors retain their scintillating ability in solution. Since fluor and sample are dissolved in the same solution, they are therefore in intimate contact with each other: the method is therefore very valuable for low-energy beta and for alpha particles, which are absorbed in the walls of the detector when placed external to a G–M tube or a scintillation crystal. Geometry is 4π, so that carbon-14 can be counted with up to 90% efficiency, and tritium with up to 40% efficiency, by internal-sample scintillation counting: the only other method to approach these figures is G–M counting of gaseous samples.

Quenching and residual phosphorescence in liquid scintillation

Like all detection methods, liquid scintillation has its problems, and one of them is quenching. This is *not* the same as the quenching of a G–M tube, and must not be confused with it: here, it means any reduction of detection efficiency by reduction of the amount of light reaching the photomultiplier. There are four causes, as follows:

1. Chemical quenching—this, the most common cause, refers to the absorption of energy from the ionising radiation without release of light photons, and arises particularly with polar compounds and dissolved oxygen.
2. Colour quenching—a coloured solution will absorb some of the photons before they reach the PM.
3. Optical quenching—this is caused by anything which reduces the transparency of the solution or the vial in which it is contained, such as the presence of two phases in the liquid or of pieces of solid matter.
4. Dilution quenching—if the fluor solution is diluted, the fluor molecules will be farther apart and the probability of interaction with them by an ionising particle is reduced (Gibson and Gale, 1967).

Residual phosphorescence of the counting vials after exposure to bright light can be troublesome: it can take up to 2 h to die away. It is eliminated by keeping the vials in the dark or in dim light before counting (Fodor–Csanyi and Levay, 1969).

Solutes and solvents in liquid scintillation

The substances and concentrations used to make up a liquid scintillation solution vary enormously, since their selection has been on a somewhat empirical basis, and the literature abounds with different recipes. The components of a solution fall into four broad categories, as follows:

1. The primary solute, the fluor—a common example is a phenyl-oxazole compound generally known as PPO, the abbreviated form of its very long name.
2. The secondary solute, or '*wave-shifter*'—the light output from the fluor may not be in the optimum wavelength range for the PM, and the wave-shifter absorbs and re-emits it at a more suitable wavelength. An example is another phenyl-oxazole, POPOP.

These solutes are dissolved at concentrations of about 5 g/l and 0·05 g/l respectively in:

3. The organic primary solvent—this forms the basic scintillation solution, an example is toluene.
4. A secondary solvent—this is sometimes necessary, because it is at once apparent that the highly organic nature of the scintillation solution raises the problem of solubility where samples are aqueous, as is the case in biological work especially. Chemical witchcraft is necessary to solve the problem of incorporating as much aqueous sample as possible in the scintillator without separation into two phases or formation of an emulsion, and without excessive quenching. The secondary solvent is both organic and water-soluble, and blends sample and scintillator together.

Examples of secondary solvents are as follows.

Ethanol or dioxan allow the incorporation of some 3% and 25% respectively of water into the scintillator. Being polar, they introduce some chemical quenching.

A quaternary ammonium compound comonly called Hyamine-10X is used primarily for the incorporation of carbon dioxide; it also allows the incorporation of relatively large volumes of water. It may be noted in passing that it will not absorb carbon monoxide, so that a mixture of the two gases can be separated into its two components and each counted separately (Houminer and Weinstein, 1968).

A detergent, Triton 100X, has come into favour recently as a secondary solvent (Fox, 1968).

A somewhat different combination of solvents is 1,2-dimethoxyethane (ethylene glycol dimethyl ether) as the primary solvent and napthalene as the secondary solvent, which is reported to dissolve up to 10% water and is recommended for low-level counting of weak beta and alpha emitters (Levin, 1962).

The efficiencies and costs of liquid scintillation mixes for aqueous tritium samples have been assessed by White (1968).

Another approach to the problem accepts that organic scintillators and aqueous samples do not readily go together and does not attempt to make them into a homogeneous mixture. The samples may be suspended into a gel along with the scintillator, or dried on to filter paper which is then placed in fluor solution (e.g. Hutchinson, 1967). Alternatively, the sample is intimately mixed with anthracene crystals and, though the mixture is not transparent, it is translucent enough not to lose much light. This method has the advantage that, because anthracene is inert and insoluble with respect to aqueous solutions, both it and the samples can easily be recovered after counting (Steinberg, 1960; Schram and Lombaert, 1962).

An entirely different and somewhat unconventional method is presented by Garrahan and Glynn (1966), who measure ^{24}Na and ^{42}K in a liquid scintillation counting system without adding any scintillant. They make use of the fact that the hard beta and gamma radiation from these isotopes excites the glass of the counting vial or the quartz face of the PM tube, and produces sufficient Cerenkov radiation* for counting with efficiencies of 26% (Na) and 62% (K) of that with liquid scintillator present; with a sodium iodide crystal (external counting) the corresponding efficiencies were 40% and 22% respectively. Johnson (1969) applies the same principle to the counting of phosphorus-32. The method is not applicable to soft radiation.

3.4 AUTORADIOGRAPHY

3.4.1 Principle

Autoradiography is based on the principle that ionising radiation has the same effect on a photographic emulsion as does visible light. A film placed in contact with a radioactive specimen will,

* Cerenkov radiation occurs when a charged particle moving in a medium has a velocity greater than the velocity of light in that medium.

when developed, be blackened according to the distribution of the activity present in the specimen, and the precise location of the activity will be revealed. Historically, autoradiography was the first method of detection of ionising radiation, for it was the means whereby radioactivity was accidentally discovered. It is sometimes known by the less pleasing name of radio-autography.

The blackening of a photographic film is described by its optical density (logarithm of the ratio of incident to transmitted light intensities). Figure 3.10 shows the relation of optical density to

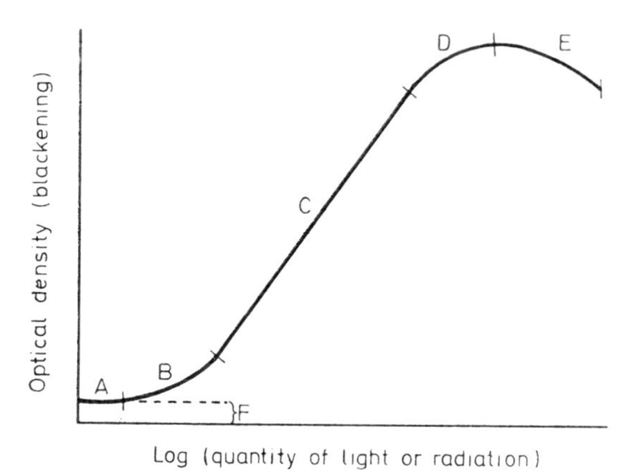

Figure 3.10. Blackening of photographic film by radiation. For explanation of lettering, see text

the amount of light (or radiation) which has struck the film. A is the region of inertia, where the film is unresponsive; B represents underexposure, and D overexposure. C is the region in which the film is normally used, in which its response is linear. E is where gross overexposure or solarisation occurs. The quantity F represents the fogging of an unexposed film; it should be less than 0·1 for a good film, and can be over 0·5 for old or badly stored film. The sensitivity of a film depends on the energy of the particles or quanta hitting it: with electrons, it is highest for those of energy up to 0·1 MeV, then declines somewhat as the energy increases to 0·4 MeV, above which it is independent of energy. Sensitivity is also affected by the development treatment applied to the film. It may be increased by using luminescent screens in conjunction with the film.

Autoradiography may be used in conjunction with gas ionisation or scintillation counting for the location of areas of activity in a large specimen (e.g. a chromatogram), these areas then being quantitatively assayed by one or another of these methods (e.g.

Vesely and Tykva, 1968). However, this can indicate only what happens to the activity, as can other methods, and the corresponding stable element may behave differently (Thomas, 1968).

3.4.2 Advantages and artefacts

Autoradiography is a very simple method, for it needs no electronics, unlike gas ionisation and scintillation counting, although it does need a darkroom for preparing and developing the films. It is an integrating process, and it may be carried on for long periods of time if low activities are being dealt with. On the other hand, this may be a decided disadvantage if a result is wanted quickly. It reveals the complete distribution of the activity in the specimen and it may thus yield more information than was expected when the experiment was designed. Autoradiography does, however, have the disadvantage of being particularly liable to artefacts, for several things other than ionising radiation can blacken the film, and 'blank' runs are often useful in helping to decide whether a particular appearance of blackening on the film is an artefact or a genuine result.

Some possible causes of artefacts are as follows:

1. Mechanical pressure—rubbing the film against hard material (grit and dust) will scratch it, and bending it will produce 'stress marks', the curved lines on films well known to all amateur photographers.
2. Light leakage—until the film is fixed after development, any light to which it is exposed will blacken it and it must therefore be kept and handled in darkness or a suitable safe-light up to this point.
3. Contamination—this can occur on cassettes and film holders by accidental spills of activity or by unnoticed pieces or dust from previous radioactive specimens.
4. Chemical reaction—this can be produced on the film by spilt chemicals or by handling it with dirty fingers.
5. Movement—of specimen with respect to film, or vice-versa, during the exposure.

Several other possible causes of artefacts are found in particular techniques. Their existence may not be suspected, and they may therefore lead the experimenter in entirely the wrong direction until they are discovered. Even then they may be troublesome and difficult to circumvent if they are inherent in the method being used. For example, in the oven-drying or press-drying of plants, chemical concentration gradients are set up as the plant loses water, and the passive movement of substances down these gradients may be misinterpreted as having happened in the living plant. This particular artefact is overcome by freeze-drying the specimens. It has been well said that skill and experience are needed to get the best out of autoradigraphy, particularly in quantitative interpretations and on the ultramicro-scale, and that it is as much an art as a science.

3.4.3 Resolution — factors affecting

Resolution in optical work (microscopy, astronomy) is defined as the minimum distance between two points which still allows them to be distinguished apart as separate entities. If the points are taken to mean point sources of radioactivity, the same definition applies in autoradiography. Several factors affect the resolution that may be attained.

Grain size of the film

Although a photographic emulsion looks uniform to the naked eye, it is seen under the microscope to be made up of silver halide crystals of varying sizes, the largest of which determines the grain size of the emulsion. The action of light or ionising radiation at any point on one crystal or grain reduces a small amount of silver halide to silver, the *latent image*, which on development catalyses the reduction of the whole grain to form the visible image, a black spot of metallic silver. The larger the crystals, the more easily does a latent image form, so that it is generally found that a fine-grain film is slower, i.e. less sensitive to light (or radiation) than a coarse-grain one; it also shows greater contrast, i.e. the blacks are blacker and the whites are whiter.

Thus, if two point sources of radiation are overlain by one grain, it will not be possible to distinguish them apart as separate sources, although it may be possible to do so if they are overlain by two separate grains. The resolving power of an emulsion can be no better than the largest grains it contains, and it is not possible to distinguish separately, or resolve, sources closer together than the grain size of the film. The finer-grained a film is, the higher is its resolving power. Besides the intrinsic nature of the film, the developer used can also affect the grain size of the emulsion and consequently the resolution.

Nature of the radiation

Since ionising particles interact with matter all along their paths (though admittedly with more intensity at the end) it seems obvious that the farther the radiation travels in the emulsion the more blackening it will produce, and a point source will not affect only the grains of the emulsion in its immediate vicinity. The extent to which the radiation penetrates in the emulsion has a marked effect on the resolution and, if two point sources are closer together than the range of the particles they emit, they will be irresolvable.

Thus alpha particles are ideal for autoradiography, for they produce a very intense localised effect along tracks only a few tens of microns long and resolution is excellent. Biologically, though, they are of interest far more as toxins and contaminants than as tracers.

With beta particles, resolution is in inverse proportion to their energy, and an isotope emitting purely soft beta particles would be the ideal choice for a biological radiotracer experiment using autoradiography as the method of detection. Tritium gives a resolution of better than 2 μm, and carbon-14 and sulphur-35 about 100 μm; with phosphorus-32, on the other hand, the resolution is worse than 3200 μm.

Gamma radiation is generally more of a liability than an asset, for it is penetrating and has a low specific ionisation, so that it causes all-over fogging of the emulsion rather than sharply defined areas of blackening.

Emulsion and specimen thicknesses and their separation

These dimensions should all be as small as possible for maximum resolution to be obtained.

Exposure time

If the film is retained in contact with the specimen too long before development, resolution will tend to be reduced in that areas of relatively high activity will be over-exposed and the fine details of the picture in them will become blacked out and lost. If the exposure time is too short, on the other hand, areas of weak activity will be under-exposed and there will be little or nothing visible in them. Determination of exposure time is somewhat empirical: it has been estimated that a fast x-ray film of the type used in gross autoradiography will be detectably blackened by some 10^5 beta particles/cm^2, and an optimal picture is given by about 10^7/cm^2. If the disintegration rate of the specimen can be roughly assessed, then knowing its area a rough estimate ('guesstimate') of the exposure time can be calculated, not forgetting that half the particles will travel away from the film. A series of exposures for different times may then be made, for (say) $\frac{1}{10}$, $\frac{1}{3}$, 1, 3, and 10 times the estimate: the shorter times will reveal the activity pattern in areas of higher activity without blacking-out of fine details, and the longer ones will bring up the areas of lower activity. If further exposures are required, they may be made using this series as a basis for calculation of the exposure time more accurately. And, if nothing else, the series will show how the exposure times must be adjusted if the initial estimate was wildly in error.

3.4.4 Techniques

Two distinct scales of operation may be distinguished in autoradiography and, along with this, two broad lines of technique. Operation on the *macro-scale* is '*contrast*' or *gross* autoradiography, in which the gross effect of many ionising particles is seen; it is carried out with, for example, chromatograms or whole plants, and usually by a *temporary-contact* technique. The film is placed in contact with the specimen only during the exposure time, and any subsequent treatment of one does not affect the other; difficulty may arise, though, in the interpretation of the results if there is any shrinkage or distortion of film or specimen during treatment. The possibility of chemical reaction between specimen and film causing an artefact can be avoided by sandwiching a thin sheet of polythene or similar material between them. X-ray films are usually used, and give quite satisfactory results; ordinary photographic materials may even be employed for simple demonstrations. The resolution is not high but, in view of the difficulty of exactly aligning specimen and film for comparison in the temporary-contact method, this is no real disadvantage.

On the *micro-scale* is '*track*' autoradiography, which looks at the tracks made in the film by individual particles. In biological work it is very often combined with a histological or histochemical staining method on tissue sections, when it is almost invariably a *permanent-contact* method. The autoradiography and the staining can be carried out on adjacent sections of a series, but it is often desirable to do both on the one section and, though this eliminates the above-mentioned difficulty of the temporary-contact method, it introduces others: these will be discussed below. Resolution becomes of great importance on the micro-scale, and the specially-designed so-called *nuclear emulsions* are employed. They are available as liquid emulsions, or as '*stripping*' *film*, a very thin material that is attached to a thick backing, for ease in handling and avoidance of mechanical damage before it is required, and from which it is easily removed or stripped off. Huxham, Lipton, and Howard (1969) discuss resolution and efficiency in electron-microscope autoradiography with radio-calcium.

Autoradiography of tissue sections

If autoradiography is carried out after staining of the tissue sections, there is a real risk that some or all of the activity may be leached out or otherwise affected by the staining treatment. If autoradiography is done before, immediately after sectioning of the tissue, this risk is removed; there is still the possibility that the preceding operations of fixation, embedding and dehydration may affect the autoradiograph, though if fresh-frozen sections are prepared by freeze-drying, the radioactivity will record on the photographic emulsion before any of the histology can affect it. However, the presence of the photographic emulsion on top of the section may now interpose a barrier to free penetration of the stains. They may well take longer to reach the tissue and, having reached it, the surplus stain may take longer to wash out, a decided disadvantage where precise timing of the staining is important; in addition the gelatine of the emulsion may itself be stained and mask the picture in the tissue.

Nevertheless, many excellent techniques for autoradiography of tissue sections have been fully worked out, and detailed instructions are to be found in the more specialised works on the subject e.g. Gude (1968). An indication of their broad principles is as follows.

'*Mounting*' *method.* The tissue sections in paraffin are floated on water, and a photographic plate (conveniently in the form of a standard microscope slide) is dipped underneath and lifted out so that the sections stick to it. After drying, the autoradiographic exposure is given, followed by dewaxing and staining. The method is fairly straightforward and easy and gives a resolution of 5–7 μm, but the photographic development may be uneven and patchy because the aqueous developer has to penetrate the wax of the section before it reaches the emulsion.

'Stripping' film method. This is the reverse of the previous method. The film is floated on water and the tissue sections, mounted on a slide, are lifted up to be covered by the film. The inert base of the film protects the emulsion from the possible chemical action of anything present in the sections. Resolution is about 2 μm and the method is widely used. The discharge of static electricity as the film is stripped from its thick backing may cause an artefact if the atmosphere is dry; this is easily overcome by breathing on the film before stripping it.

Coating method. This uses a liquid emulsion that is painted on to the sections and allowed to dry. It is essentially the same as the stripping-film technique, but it presents more technical difficulty in handling the emulsion.

Molecular autoradiography. This is a high-resolution method which has been developed for use in electron microscopy and in which a monolayer of silver halide crystals is deposited on an ultra-thin tissue section. Its resolution is an order of magnitude better than the previous methods, going down to 0·1 μm and it is a powerful tool for the study of cellular ultrastructure.

Recently, the registration of charged particles in solids has been proposed as an alternative to autoradiography in the life sciences (Hamilton, 1968), and evaporated silver bromide has been suggested as an ion and particle detector (Masters, 1969).

3.5 ARRANGEMENT OF COUNTING CIRCUITRY

The electronic gear in a radio-isotope counting room may well appear frighteningly complex to the uninitiated: a vast number of switches, knobs and dials in bewildering array that he hardly dare touch for fear of damaging something. But it may be broken down into a small number of basic units whose functions are quite simple, and which can easily be represented in the form of a '*block diagram*'. Once these basic units have been identified in a counting assembly, its operation will be greatly simplified and much more easily understood.

An ionisation chamber or scintillation counter detector must be connected to a suitable circuit both to drive it and to count the pulses it produces. For a G–M tube, the basic circuit (Figure 3.11) comprises the following:

1. An adjustable high-voltage supply—this powers the tube.
2. A quenching pre-amplifier or 'probe unit'—this applies a fixed

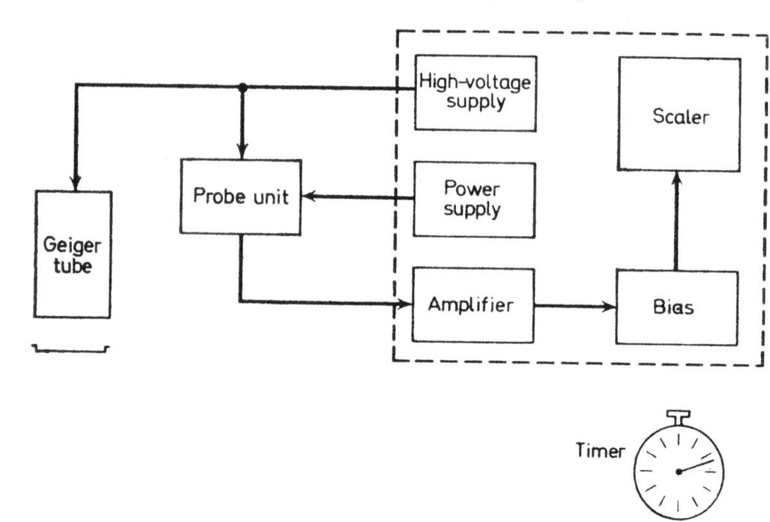

Figure 3.11. Block diagram of simple G–M *counting assembly*

quenching time to the tube; the counts are fed into it via the high-voltage supply lead.

3. A power supply—this drives the probe unit, and is usually integrated with the high-voltage supply.
4. An amplifier—this accept the pulses from the probe unit and passes them into the scaler.
5. A scaler—this counts the pulses: the counts are typically displayed on a bank of discharge tubes, commonly called '*dekatrons*', which present a glow discharge in one of ten positions. Older machines use a mechanical register to record hundreds and above, and the latest ones give a digital readout on an illuminated screen. The scaler contains a facility for counting mains frequency while it is warming up, to test that the dekatrons are working properly.
6. A discriminator bias (or back bias) is invariably incorporated in the scaler. The function of this bias is to prevent pulses below its setting being passed on to the scaler and counter. Its lowest setting is fixed to cut out electronic 'noise' in the preceding circuitry. The casing holding the scaler may also incorporate items (1), (3) and (4).
7. A timer is obviously essential to obtain count rates, the number of pulses in a certain time; in its simplest form, it is a stop-clock or stop-watch worked simultaneously with the start/stop switch on the scaler. A refinement is an automatic timer, either electronic or mechanical, which stops the scaler after

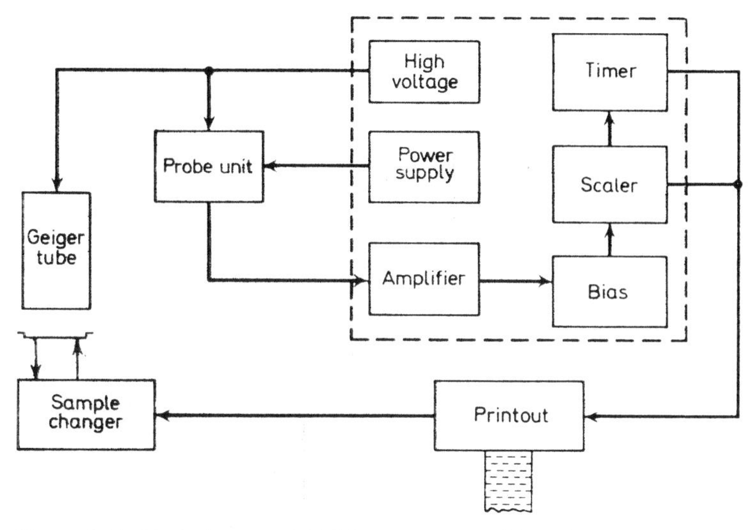

Figure 3.12. Block diagram of G–M *counting assembly with automatic sample changer*

a preset time, or is stopped after a preset number of counts. It too may be built in with the scaler in one unit.

8. Addition of an automatic sample changer (Figure 3.12) enables the counting assembly to operate entirely unattended, with a consequent saving in man hours. The changer comes into action after a preset time has elapsed, when it:
   (a) prints out the number of counts recorded in that time;
   (b) removes the sample from the detector and replaces it with the next one to be counted; and
   (c) resets the scaler and the timer to zero and restarts them.

   Alternatively, the automatic sample changer may operate after a preset number of counts has been recorded, when it prints out the time taken to record them ('time for unit events' as opposed to 'events in unit time'). A weakly active sample may not be able to give this number in a reasonable time and, to avoid spending far too long on it, an over-riding time control is fitted, which changes the sample at the end of a preset time period whether or not it has accumulated the preset number of counts.

In scintillation counting, the probe unit is not required, and an additional first-stage amplifier is inserted between the photo-multiplier PM and the scaler. There are two discriminators, as outlined in Section 3.3.3, forming the adjustable *gate* or *window*.

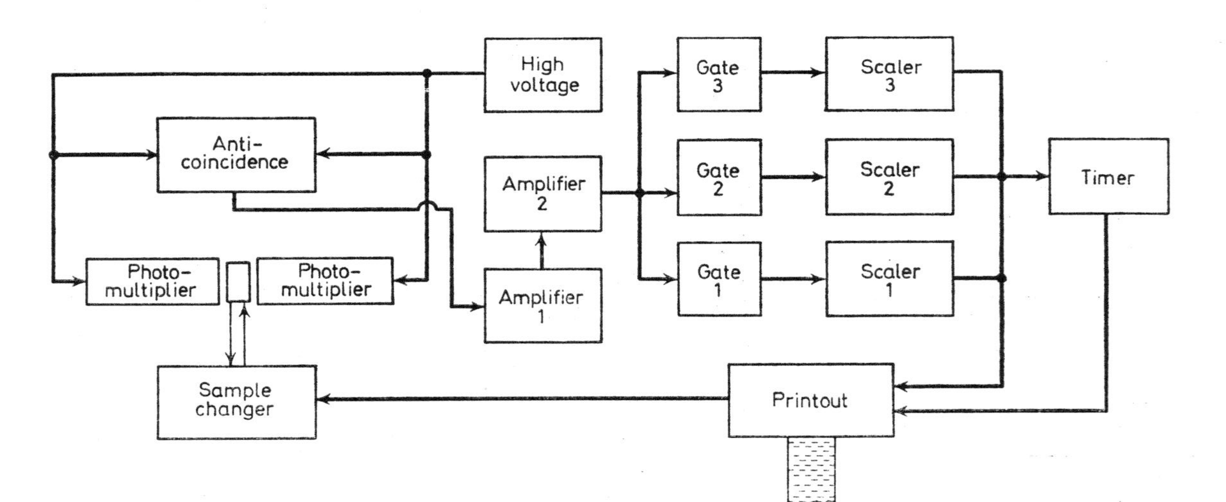

Figure 3.13. Block diagram of multi-channel anti-coincidence scintillation counter

The more advanced machines may have three such channels, all independently adjustable, and each with its own scaler linked to the timer and printout; here the sample changer is invariably set to operate after a preset time.

Anti-coincidence circuitry reduces background: two PMs in contact with the same fluor are connected so that a count is recorded only if it is detected by both PMs at the same time, as it will be if it has arisen in the fluor, whereas 'noise' pulses arising independently in the PMs will coincide and be recorded only occasionally by chance. Anti-coincidence circuitry has another advantage if the coincident output pulses from the PMs can be summed; this shifts the spectrum towards higher energy levels, and allows a better separation of sample pulses from any noise that is not eliminated by the anti-coincidence. The block diagram of a multi-channel anti-coincidence scintillation counter is shown in Figure 3.13.

3.6 SEMICONDUCTOR DETECTORS OF IONISING RADIATION

The use of semiconductors as detectors of ionising radiation was theoretically possible for some forty years before material could be obtained pure enough to allow usable devices to be made, but within the past ten years or so they have developed rapidly and are now commercially available. They have several advantages over scintillation and gas ionisation counters, being compact, relatively cheap, and having excellent energy resolution. They need but a simple power supply since the applied voltage required to drive them is not high, 10–30 V only.

In principle, semiconductor detectors are solid-state versions of gas ionisation detectors; they are based on either germanium or silicon. Ionising radiation produces electron–'positive-hole' pairs (charge carriers) which are swept to the outside of the detector by the applied potential, where they produce a measurable pulse. In silicon, 3·5 eV are needed to form a pair, and in germanium 2·9 eV, so that about 10 times more charge is produced than in gas ionisation and the energy resolution is consequently about three times better, being proportional to the square root of the number of ion pairs formed. In addition, the resolving time is shorter. Reduction in efficiency and resolution arises from polarisation: localised imperfections in the crystal trap the charge carrier to produce an electric field opposing the applied one. Polarisation is lower the higher the resistance of the crystal, as the number of imperfections is lower. Random thermal motion of the atoms in the crystal generates electron–hole pairs, producing a fluctuating background (leakage current or dark current) which would swamp the signal from the detected radiation if it were not reduced. Leakage current is greater in germanium than in silicon, because the gap between the valence and conduction bands in its atom is less. High resistance lowers leakage current (besides decreasing polarisation).

Surface barrier detectors

There are various forms of semiconductor detector, of different characteristics, in both germanium and silicon; they include the very similar surface barrier and surface junction types, and the p–i–n or positive–intrinsic–negative type. In the surface barrier detector, a thin p-type surface is formed on one face of a wafer of n-type material (or vice versa), giving a dipole region negative on one side and positive on the other; application of an external voltage creates a charge depletion region in between, of very high resistance and free of charge carriers and polarisation centres, or in effect a pair of capacitor plates with a high-resistance semiconductor in between as the dielectric. Gold and aluminium are evaporated on to the p- and n-type surfaces respectively to make the external connections.

Surface barrier detectors have very low sensitivity (or are unresponsive) to gamma radiation and to neutrons, so that they have a distinct advantage over scintillation and gas ionisation counting for alpha radiation in the presence of gamma. An example of this is in health physics monitoring for such things as radium, thorium and plutonium; normally the amount of activity is very small, which necessitates accurate low-level counting of the alpha radiation.

Lithium-drifted detectors

The depth of the sensitive layer is restricted by the resistance of the material, which is determined by the number of p- and n-type impurities. It is made greater in the p–i–n detector either by using ultra-high purity material or by very careful control of the amount of impurity. Most of the n-type impurities are removable in the purification process but some p-type always remain; re-introduction of n-type impurity in the exactly equivalent amount and attachment of each n-type atom to a p-type impurity will 'compensate' the material and give it very high resistance. Lithium is suitable, being a small sized atom; it is applied to one face of the material, and is diffused or drifted into it by the application of a reverse bias (voltage) in the Pell process, to produce lithium-drifted detectors. Their size is limited by the quality of the starting material and is normally about 60 mm square and 6 mm deep.

These detectors are usable for gamma radiation; for energies above 100 keV those based on germanium rather than silicon are used, because the higher atomic number gives greater stopping power for photo- and Compton electrons. They cannot be kept at room temperature for more than an hour or so, because of mobility of the lithium atoms, and they are stored in dry ice (solid carbon dioxide). They are used in liquid nitrogen, when their resolution is ideally 2 keV FWHM (Section 3.3.3) and 4–7 keV under usual laboratory conditions: being so good, this may remove the need for chemical separation of mixtures of gamma emitters before spectrometry.

A range of lithium-drifted silicon detectors is available commercially in which lithium compensation is described as being stabilised by a 'diffused window' of silicon 50 $\mu g/cm^2$ thick. This is rugged enough to be cleaned, should it become contaminated, and the detector is contrasted to the fragile surface barrier pattern which cannot be handled or cleaned. They can be stored at room temperature, and one design operates between $-30°C$ and ambient, so that it would appear to be more generally useful than those needing to be used at low temperatures. Its resolution is 50 keV FWHM or better, depending on the temperature of operation.

References to semiconductor detectors are: Bilaniuk (1962), Friedland (1965), Haschizume, Legrand, and Hareux (1965), Perry (1965), and Cooper *et al.* (1968).

REFERENCES

BILANIUK, O-M., (1962). 'Semiconductor Particle Detectors', *Scient. Am.*, **207**, October, 78–92.

COOPER, J. A., WOGMAN, N. A., PALMER, H. E., and PERKINS, R. W., (1968). 'Application of Solid State Detectors to Environmental and Biological Problems', *Hlth Phys.*, **15**, 419–433.

FODOR-CSANYI, P., and LEVAY, B., (1969). 'Phosphorescence and Liquid Scintillation', *Int. J. appl. Radiat. Isotopes*, **20**, 223–228.

FOX, B. W., (1968). 'Application of Triton X100 Colloid Scintillation Counting in Biochemistry', *Int. J. appl. Radiat. Isotopes*, **19**, 717–730.

FRIEDLAND, S. S., (1965). 'Gamma-ray Assay with Lithium-Drifted Germanium Semiconductor Nuclear Detectors', *Proceedings of the Symposium on Radio-Isotope Sample Measurement Techniques in Biology and Medicine*, International Atomic Energy Agency, Vienna, 635–645.

GARRAHAN, P. J., and GLYNN, I. M., (1966). 'Measurement of ^{24}Na and ^{42}K with a Liquid Scintillation Counting System Without Added Scintillant', *J. Physiol., Lond.*, **186,** 55P–56P.

GIBSON, J. A. B., and GALE, H. J., (1967). 'Fundamental Approach to Quenching in Liquid Scintillators', *Int. J. appl. Radiat. Isotopes*, **18,** 681–688.

GUDE, W. D., (1968). *Autoradiographical Techniques*, Prentice-Hall, New York, 113 pp.

HAMILTON, E. I., (1968). 'Registration of Charged Particles in Solids', *Int. J. appl. Radiat. Isotopes*, **19,** 159–160.

HASCHIZUME, A., LEGRAND, J., and HAREUX, F., (1965). 'Spectrometric Analysis of Beta Radiation with Semiconductors', *Proceedings of the Symposium on Radio-Isotope Sample Measurement Techniques in Biology and Medicine*, International Atomic Energy Agency, Vienna, 669–684.

HOLLANDER, J. M., and PERLMAN, I., (1966). 'The Semiconductor Revolution in Nuclear Radiation Counting', *Science, N. Y.*, **154,** 84–93.

HOUMINER, Y., and WEINSTEIN, M., (1968). 'Simple Technique for Collecting $^{14}CO_2$ for Liquid Scintillation Counting', *Int. J. appl. Radiat. Isotopes*, **19,** 663–664.

HUTCHINSON, F., (1967). 'Assay of ^{45}Ca in Biological Materials by Liquid Scintillation Counting', *Int. J. appl. Radiat. Isotopes*, **18,** 136–137.

HUXHAM, G. J., LIPTON, A., and HOWARD, B. M., (1969). 'Resolution and Efficiency in Electron-Microscope Autoradiography with Radioactive Calcium', *Aust. J. exp. Biol. med. Sci.*, **47,** 299–304.

IAEA, (1965). *Proceedings of the Symposium on Radio-Isotope Sample Measurement Techniques in Biology and Medicine*, International Atomic Energy Agency, Vienna, 427, 635–645, 669–699.

JOHNSON, M. K., (1969). 'Counting of Cerenkov Radiation from Phosphorus-32', *Analyt. Biochem.*, **29,** 348.

LEVIN, L., (1962). 'Liquid Scintillation Method for Measuring Low Level Radioactivity of Aqueous Solutions', *Analyt. Chem.*, **34,** 1402–1406.

MASTERS, J. I., (1969). 'Evaporated Silver Bromide as an Ion and Particle Detector', *Nature, Lond.*, **223,** 611–613.

PERRY, K. E. G., (1965). 'Some Applications of Surface Barrier Solid-State Detectors for Alpha Activity', *Proceedings of the Symposium on Radio-Isotope Sample Measurement Techniques in Biology and Medicine*, International Atomic Energy Agency, Vienna, 685–699.

ROBERTS, W. A., (1967). 'A New Performance Criterion for Liquid Scintillation Spectrometry', *Bio-med. Engng (G. B.)*, May.

ROSEN, J. C., LAUREE, G. R., and EISENBUD, M., (1967). 'Carbon-14 and Tritium Measurement by Bremsstrahlung', *Science, N. Y.*, **157,** 77–78.

SCHRAM, E., and LOMBAERT, R., (1962). 'Determination of Tritium and Carbon-14 in Aqueous Solution with Anthracene Powder', *Analyt. Biochem.*, **3,** 68–74.

STEINBERG, D., (1960). 'New Approach to Radio-Assay of Aqueous Solutions in the Liquid Scintillation Spectrometer', *Analyt. Biochem.*, **1,** 23–39.

THOMAS, W. A., (1968). 'Calcium Distribution in Leaves—Autoradiography Versus Chemical Analysis', *Int. J. appl. Radiat. Isotopes*, **19,** 544–545.

VESELY, J., and TYKVA, R., (1968). 'Correlation of Simultaneous ^{3}H and ^{14}C Autoradiography with Gas Counting', *Int. J. appl. Radiat. Isotopes*, **19,** 705–706.

WHITE, D. R., (1968). 'Assessment of the Efficiencies and Costs of Liquid Scintillation Mixes for Aqueous Tritium Samples', *Int. J. appl. Radiat. Isotopes*, **19,** 49–61.

See also the books referred to at the end of Chapter 4.

Chapter 4

THE USAGE OF ISOTOPES

4.1 PRINCIPLES OF THE ISOTOPE TRACER TECHNIQUE

4.1.1 Definition of a tracer

Something of the definition of a tracer is implicit in the name itself. It is a means which enables something to be located wherever it happens to be, or traced, more easily than would be possible otherwise. An example from everyday life is seen in the handling of assorted packages; those requiring special handling are given a special label so that they can be readily picked out from the remainder. Biological examples are the 'ringing' of birds, an indispensable aid in the study of migration, and the use of heavy metals in electron microscopy, where they give their peculiar image that greatly aids in the interpretation of the final picture. A distinctive atom in a molecule enables the metabolic changes of the compound to be followed, as for example with arsenical or antimonial drugs; any other compounds that are found to contain this atom will almost certainly be metabolites of the original compound. Such compounds are referred to as being *labelled* or *tagged.* If the compound does not contain an atom or group able to act as a label, then one may be added, as with the attachment of a benzene ring to fatty acids to assist study of their metabolism, this group not being present in natural fatty acids. However, this will reveal the metabolism of benzenoid fatty acids rather than that of natural ones and, though it is highly likely, there is at the same time no guarantee that the two will be the same.

Obviously a tracer must be easily recognisable, and it must therefore be something that is not normally found in the system being studied, unless by a fortunate chance an unusual feature happens to occur naturally and form a label ready-made. It is also fairly obvious that the tracer should not then interfere with the normal working of the system or behave differently from what it labels, but the very fact that it is not normally present may well cause this to happen. For this reason the tracer technique was very restricted in its use and not of general biological application in the pre-radio-isotope era, but the coming of radio-isotopes revolutionised it, and enabled it to spread into practically all branches of biology, particularly metabolic studies. The reason for this is that radio-isotopes fulfil the definition of an ideal label as nearly as it is possible for anything to do: they are to all intents and purposes identical to the naturally occurring stable element in so far as their chemical behaviour is concerned (but see Section 4.2.1) but they differ in that they emit radiation whereby they can be identified and quantified with great ease. Compounds labelled with a radioactive atom are commonly called hot, in view of the heat-producing property of a radio-isotope like polonium. The introduction of an isotope into an experiment at one point in time is often referred to as *pulse-labelling* or *spiking*.

Tracer theory is discussed by Rescigno (1968).

4.1.2 When to use isotopes

Radio-isotopes should not be used in every experiment, or just for the sake of it for, though they are very powerful tools, they do have their limitations, and there is an element of danger attached to them. Their indiscriminate and hasty use, without proper forethought and care, can easily lead to wastage of time and material and introduce serious hazard into the experiment.

Two questions should be asked before the decision is taken to use isotopes in an experiment:

1. Will the results be at least as informative and relevant as those that would be obtained by non-isotopic methods?
2. Would the results be significantly more difficult to obtain by non-isotopic methods?

If the answers to these two questions are 'no', then the use of isotopes in the experiment is not justified.

4.1.3 Requirements of a tracer

The basic requirements to be fulfilled are that it must be possible to label the compound being studied, first, with a suitable isotope, and secondly, in a suitable position.

A suitable isotope is one whose half-life is long enough to enable it to last for the duration of the experiment. Usually this presents no difficulty, but there are exceptions, as in a long-term study of the effect of fluorine on tooth decay (dental caries), for the only radio-isotope of fluorine is ^{18}F whose half-life is just 1 h 51 min. In addition, the radiation properties of the isotope should be satisfactory: thus a powerful beta or gamma emitter is of little use for micro-autoradiography, and poorly penetrating radiation is difficult to detect, a constantly recurring problem with tritium and carbon-14. Stable isotopes overcome the difficulty of short half-life (if there is one for the element of interest), but their detection and quantitation, involving, for example, mass spectrometry, is considerably more difficult than with radioactive isotopes.

A suitable position for the label is one in which it will provide as much information as possible about the molecule or the part that is of interest. The choice of position is restricted to one or at most a small number for practically all elements except carbon and hydrogen. The label must be firmly attached, otherwise the only information gained from the experiment will be that the compound has lost its label, which is not particularly helpful! This is a problem with hydrogen isotopes, (D and T), for hydrogen is labile and freely exchangeable not only in the obvious positions such as —COOH, —OH, and —NH_2 groups but also in others too, from which the label will quickly spread to a large number of compounds that have no relation whatever to the metabolism of the one that was labelled originally, giving completely meaningless results.

Tracer nomenclature

There is as yet no internationally-recognised system for naming isotopic compounds, but the name must specify the compound without ambiguity. The position(s) of the isotope must be shown if there is more than one possibility: thus the name glucose-^{14}C is incomplete and needs qualification with the position of the isotope, e.g. glucose-l-^{14}C if the label is on the carbon atom in position 1. By contrast, no indication of position is necessary with methanol-^{14}C. A capital U following the name of a carbon-14 compound denotes statistically uniform labelling: e.g. glucose-^{14}C (U) is glucose in which all six carbons have the same specific activity. Similarly, a G after the name of a tritium compounds means that the label is distributed generally throughout the molecule, although not necessarily uniformly.

Tracer synthesis

A requirement for a particular pattern of labelling raises the question of synthesis. Labelled compounds may be made either '*in vitro*' i.e. by chemical methods or enzymatic reactions, or '*in vivo*' i.e. by making use of the versatile metabolic capabilities of micro-organisms, bacteria in particular, to carry out reactions that would be extremely difficult or impossible otherwise; biosynthetic methods are known as '*isotope farming*'. Activation by irradiation with neutrons (Section 4.6.4) is sometimes applicable. It is true to say that in theory it is possible to label almost any compound satisfactorily in almost any position, although practicability is another matter on the grounds of expense and time involved in the preparation; the theoretical ideal may turn out to be unobtainable in practice, and may have to be forsaken for something else.

4.2 VALIDITY OF RADIOTRACER EXPERIMENTS

It is, of course, assumed that radiotracer experiments, like any others, give a true and valid result and, in making this overall assumption, various other assumptions are made concerning the purity and behaviour of the tracer and its effects on the system it is labelling. These are discussed in turn below.

4.2.1 The isotope effect (Simon and Palm, 1966)

It is assumed that the radio-isotope behaves in exactly the same way as the stable element, so that the labelled compound has the same biological properties and metabolic fate as the unlabelled one. Strictly speaking, this is not true, for there are differences in chemical behaviour between stable and radioactive nuclides of an element, referred to as isotope effects. Isotope effects are quantitative rather than qualitative, and derive from differences in the atomic weights of the isotopes, the strength of the bond between two atoms being proportional to the square root of the masses involved. Isotope effects are thus obviously greatest with the lightest elements, particularly with hydrogen and tritium whose masses are in the ratio 1 : 3, the highest for any pair of isotopes. Deuterium also shows differences in behaviour from hydrogen, for example, mice cannot survive if given pure D_2O instead of ordinary water (Smith, 1968). In spite of such isotope effects though, D and T can still be used successfully as tracers for H. The magnitude of the isotope effect falls off rapidly with increasing atomic number and, in biological work at least, is almost certainly within the limits of experimental error and of no significance above $Z = 6$ (carbon).

The isotope effect may be seen in two ways, *intra-* and *intermolecularly*. The first is found when the molecule contains two identical groups of which only one is labelled, e.g. malonic acid-1-^{14}C, $^{14}COOH$-$^{12}CH_2$-$^{12}COOH$. Decarboxylation involves breaking bonds between carbon-14 and carbon-12 in the labelled acid and between two carbon-12s in the unlabelled one; the former bond is slightly stronger, so that when the decarboxylation reaction goes to completion, slightly less CO_2 and slightly more CH_3COOH will be formed from the labelled than from the unlabelled acid. By contrast, the more common intermolecular isotope effect is exemplified by the decarboxylation of carboxyl-labelled benzoic acid; the rate of CO_2 formation from the labelled acid, $C_6H_5{}^{14}COOH$, will be slightly slower than from the normal acid, $C_6H_5{}^{12}COOH$, but the specific activity of the carbon dioxide at completion of the reaction will be the same as that of the original benzoic acid. Thus isotope effects can be ignored unless they are intramolecular, or rates of reaction are being studied.

4.2.2 Purity of the labelled material

This is a more complex matter than with inactive chemicals, first, because it is difficult to purify the minute quantities involved without loss; and secondly, radioactive materials are liable to decompose under the influence of their own radiation, by self-irradiation, so that they are inherently less stable than their non-radioactive counterparts. The purity of a radiochemical should always be checked before use, especially if there is any doubt, otherwise a completely false picture of the distribution and fate of the tracer may be obtained.

The standards of purity required of radiochemicals vary greatly according to the needs of the experiment, and may be considered from three aspects, as follows:

1. Chemical purity—this has its usual meaning. It may be noted that solutions of carrier-free nuclides, though highly radioactive, can easily pass the chemical tests for AR distilled water, so that chemical purity is of little relevance here.
2. Radionuclidic purity—this refers to the amount of the total radioactivity present which is in the form of the stated nuclide.

Radionuclidic impurities, at high specific activity and very low chemical concentration, may arise through side reactions in the nuclear reactor when the nuclide is being made, or from the presence of impurities in the target. They can be minimised by using the purest possible materials in the target, and by careful control of the irradiation conditions and the subsequent processing of the preparation.

3. Radiochemical purity—this states how much activity is in the stated chemical form, taking note of configuration and the position of the label.

Decomposition of radiochemicals

This can take place through four effects, three of which are related to their radioactivity and one which is not.

The internal radiation effect. This is the alteration of one of the atoms in a molecule simply because that atom is radioactive and transmutes into another element. It produces radiochemical impurities only if there is more than one radioactive atom per molecule.

The primary external radiation effect. This is the alteration of a molecule by interaction with a particle or gamma photon emitted from a radioactive atom.

The secondary external radiation effect. This is the attack on a molecule by a reactive ion or molecule (e.g. a free radical) produced by the above primary effects. Radiochemical impurities are produced by these external effects; again, they may have very low chemical concentration coupled with very high specific activity. A coloured preparation need not necessarily be very impure, for very few free radicals may be needed to give an intense colour.

The susceptibility of the compound to radiation breakdown is usually expressed as the *G(–M) value*, the number of molecules irreversibly altered per 100 eV of energy absorbed. It may be noted that many aromatic compounds have low $G(-M)$ values and are relatively stable to radiation attack. $G(-M)$ values can vary quite widely with conditions of storage, etc., and information on the factors affecting them is gradually being accumulated.

Radiation decomposition is a problem encountered more during storage than during the actual course of an experiment, whose time-scale is usually short. Such decomposition may be limited by several means, such as storage at low temperature, avoiding frequent thawing, and using the material as soon as possible after purification or preparation. Separating the radioactive molecules also helps, as it enables the radiation to escape without causing damage. Thus dilution with inactive carrier is effective, though it reduces the specific activity irreversibly and this may be diametrically opposed to the needs of the experiment. The substance may be spread over a large area, e.g. on filter-paper. Alternatively, it may be dissolved, or diluted if it is already in solution, although care must be taken in the choice of solvent, because free-radical chain reactions may be set up which will do more harm than good. A solution may also be 'diluted' with an inert solid such as glass beads.

Chemical effects. These are the final factor in the decomposition of radiochemicals. They are not connected with the radioactivity of the material, and take place in inactive material also; they become of significance with active preparations simply because the chemical concentrations are extremely small: e.g. a solution containing 1 pCi/ml of carbon-14 at a specific activity of 0·1 mCi/mg is about 10^{-6} mol/l. The behaviour of these very small masses and dilute solutions sometimes differs unexpectedly from that of macroscopic amounts, but it is 'abnormal' only because it is not significant with the latter. In addition, the sensitivity of isotopic methods reveals behaviour which would otherwise go undetected. Thus, adsorption from a solution on to the surface of the containing vessel is quite unimportant in a volumetric titration with decinormal solutions, but it may involve a significant proportion of the whole amount present in a 10^{-6} mol/l solution. (However, this can be put to good effect in the use of crystals and precipitates as adsorbents to de-activate large volumes of weakly radioactive waste.) Similarly, a compound which forms a precipitate at normal chemical concentrations may behave colloidally at low concentrations, loss by evaporation may occur with compounds that are normally taken to be non-volatile, such as glycerol; and susceptibility to oxidation or reduction may be greatly increased.

4.2.3 Radiation effects

The possibility of radiation injury must be guarded against. Exposure of the experimenter to ionising radiation is an obvious hazard, though it will not invalidate the experiment; it will be controlled and checked by observance of the laboratory rules, calculation of possible doses, and other safety precautions (Section 6.6). A less obvious risk arises from the possibility that the system under study may be altered, particularly through self-irradiation of the labelled compound, which has been dealt with above. Radiation damage to the rest of the system is unlikely except in long term experiments with intact animals or plants and high doses, or where an isotope becomes concentrated in a particular region of the body, as with the thyroid gland and iodine. However, most experiments are short-term and are over before any radiation damage that may have occurred can manifest itself.

4.2.4 Physiological conditions

Finally, it is assumed that conditions in a tracer experiment are physiological, including the chemical concentrations of the various compounds present, and that addition of the tracer will not alter them. The intermediates in a sequence of reactions normally exist transiently and are present only in very small amounts. If such a compound is used as a tracer, it is possible that even the minute chemical quantity added may increase the total concentration beyond the normal physiological range, and the experiment will then become unphysiological.

4.3 ADVANTAGES OF THE TRACER TECHNIQUE

Perhaps the most important advantage is that tracers enable experiments to be done under physiological conditions in the intact animal or plant (subject to the conditions above being fulfilled) to a far greater extent than with any other technique. The use of tracers in biology is not restricted to *in vitro* experiments in the laboratory, and is a most valuable and useful tool in applied work in the field: for example, small animals may be tagged with radioisotopes and their movements followed thereby (Gerrard, 1969).

Very small amounts of isotope are required, as the sensitivity of detection and determination is many times that of chemical methods, and in fact the quantities used are so small that they cannot be detected by any means other than by their radiation. For example, 1 mCi of carbon-14 weighs 0·218 mg but gives $3{\cdot}7\times10^{7}$ disintegrations per second, and a very small fraction of this, of the order of 10^{-4} μg, which gives about 20 disintegrations per second, can be detected with no difficulty. Thus, in theory, 1 mCi of carbon-14 could provide enough material for about 10^{6} determinations so that, although the initial purchase of the isotope may be expensive, each experiment may be relatively cheap.

The tracer method is a physical method of determination, and it shares the advantages that most physical methods have over chemical ones. It does not destroy the thing it determines, so that determinations can easily be repeated on the same sample without having to repeat the whole experiment. The preparation of samples is greatly simplified or unnecessary, for the activity can be determined where chemical separation would be difficult or impossible. Where chemical purification is required, the samples can be checked quickly to see if the activity is high enough to warrant further working up—if not, they can be discarded without wasting time on them and finding nothing at the end.

Isotopes enable one to find out what would be impossible otherwise: the precursors and fate of all the atoms in a molecule, for example and, by double labelling (Section 4.6.1), to distinguish between two sources of one element, such as ferrous and ferric ions for the formation of haemoglobin, or between molecules formed before and after a certain point in time.

Disadvantages and limitations are relatively few, and though they do not seriously reduce the value of the technique, they must be appreciated. Specialised equipment is needed, except for purely qualitative autoradiography and, along with it, a competent electronics technician to maintain it. All this makes for expense. Also, it is only the label which is followed, and isotope methods can give no direct information about the parts of the molecule that are unlabelled. Finally, there may not be a suitable isotope (Section 4.1.3).

4.4 EXPERIMENTAL DESIGN

4.4.1 The need for planning

It goes almost without saying that all experiments should be planned, but isotope work demands a particularly high standard of planning and execution if it is to yield the best results, in view of the potential hazard from radio-isotopes (Section 6.1). Accurate and precise work is entailed in the dispensing and handling of small amounts of radioactive materials, and the apparatus must be clean not only in the ordinary or chemical sense of the term, but also radio-clean, i.e. free from contamination with isotope. It may be necessary, too, to sterilise it where it would not be necessary on the macro-scale, so as to eliminate decomposition of the tracer through microbiological attack. The experiment should be thought out beforehand all the way through and the 'look-see' attitude ('chuck in a bit of isotope and see what happens') is quite out of place.

Several factors come into the reckoning when an isotope experiment is being planned, and they all have a bearing on its design.

To some extent they interact with each other, a decision in one direction affecting the decision in another, and it is not possible to say which is the factor of primary importance in experimental design, for this varies with circumstances.

4.4.2 Type of equipment and form and quantity of sample

An experiment may have to be planned around a limited range of counting gear, for electronic equipment tends to be expensive and, since it cannot usually be bought *ad lib.*, it is not always possible to realise the theoretical ideal of exactly the right equipment to suit the experiment. If there is no choice in the type of equipment available, it will determine both the type and energy of radiation that can be used and the form which the samples must finally take for counting. If there is a choice of detectors, these same two factors will determine which to employ; each type of radiation has a method of detection which is best suited to it (the detector of choice), others that are less suitable, and some methods that cannot be used. Thus external-sample scintillation is most satisfactory for gamma, and G–M counting for hard beta; soft beta is best counted by a proportional counter, windowless G–M tube, or internal-sample liquid scintillation, but cannot be detected very efficiently, if at all, with a liquid G–M counter. Which method of detection is chosen will determine the form of the sample. Alternatively, if there is more than one detector available and suitable for the radiation being used, the choice of detector may be decided by three factors: first, the physical form of the samples, whether they can be conveniently converted to another form and whether they are to be counted *in vitro* or are present within the body of an animal or plant *in vivo;* secondly, how much sample there is; and thirdly, the concentration of the radioactive element in the samples. The guiding principle is the attainment of the maximum overall counting efficiency; this is perhaps expressed most simply by saying that the product of the following three quantities should be maximised: the amount of sample that the detector can handle or the amount available, whichever is the smaller; the efficiency of the detector for the type and energy of radiation being used; and the concentration of radioactive element in the sample.

Finally, the experimenter must decide what is to be measured and how to measure it, so that the results will be amenable to statistical analysis and yield as much statistically reliable information as possible. If this is not done, it may be found that an important measurement has been omitted.

4.4.3 Choice of isotope and labelled compound

It would simplify matters if it were possible to have the half-life and radiation properties of an isotope made to order, but this is of course not so, and the choice of which isotope to use is often Hobson's: there is just the one isotope, and that is that. Sometimes there is a choice: it may be between a stable and a radioactive isotope, when the latter is nearly always more useful; or between two active ones, when the respective radiation properties and half-lives must be considered, and also the maximum available specific activities. In many cases, the exact position of the label in the molecule is not critical (subject to being firmly attached) and the cheapest available compound can be used; it is not always true that a U- or G-labelled compound is cheaper than one labelled in a particular position. If the desired compound is not available or is very expensive, it may be possible to make it in one's own laboratory from a cheaper precursor. The suppliers can be consulted for advice and information about compounds not in the catalogues, and will give what help they can. Labelling with an isotope of the same element, *isotopic labelling*, is not always possible, and an isotope of a different element may have to be used as a tracer, e.g. in the iodination of proteins; this is *foreign* or *non-isotopic labelling*.

The price of the available isotopes and labelled compounds is a consideration for, though radiochemicals can be quite cheap, some are very expensive or unobtainable because of difficulties in production of the isotope and in synthesis of the labelled molecule. Many isotopes are made by neutron irradiation (Section 2.4.4), and the higher the neutron capture cross-section of a nucleus, the easier it is to convert it to another by neutron irradiation. For example, nitrogen-14 has a low cross-section for the (n, p) nuclear reaction by which carbon-14 is made, and carbon-14 compounds are therefore expensive compared with many other reactor-produced isotopes, for large amounts of starting material must be exposed to high fluxes for long periods. In addition, carbon-14 compounds are often required with high specific activities, which increases their cost significantly.

4.4.4 Scale of operations

The scale of operations governs the quantity of isotope to be used, and should be kept as small as possible for, as seen above, there is an increasing danger as the amounts are increased of radiation damage to the system, hazard to the experimenter, and unphysiological chemical concentration of isotope, apart from greater cost. The quantities should not be so small, however, as to present difficulty in handling; there is no point in trying to work on the ultra-micro scale if one is not geared up (or down) to it, for it will only result in gross experimental error and completely unreliable results. The guiding principle should be, not to 'put in plenty of isotope so as to be on the safe side' (which can be a contradiction in terms), but rather to use the minimum possible amount of activity consistent with reliable results.

The extent of dilution of the activity in the experiment has to be estimated, and enough activity must be put in initially to make the final samples that are taken for counting sufficiently active, for it is obviously useless to try and assay them if they are indistinguishable from background. Dilution increases with the size of the organism or the experimental system, and any advantages conferred by large animals may well be offset by this greater dilution. Calculation of the dilution is approximate at the best, and can often only be a guess, for one of the several factors involved may well be the very thing one is trying to find out. It is a wise precaution to count samples from the several different compartments of the experiment at successive times so that, if the calculation was in error and the activity has not gone in the expected or hoped-for direction, it will at least be possible to find out where it has gone, and plan a better experiment next time. It is usual to allow a margin for errors in calculation or for any unforeseen factors, and to put in rather more activity than might strictly appear necessary at first, cutting down the quantity as the experiment is refined.

How much activity is required in the samples will depend on the desired precision of the experiment; if the counting error is to be 1% the activity should be such as to give 10^4 counts in a reasonable time (Section 4.5.5). Less counts are needed, for example, to distinguish partial or complete utilisation of an administered compound than to determine the exact level of uptake.

4.4.5 Reproducibility of results

Reproducibility is important in all quantitative experimentation, and may influence the choice of method, since consistent results are more easily obtained with some techniques than with others. Some methods, particularly liquid G–M counting, present the risk of residual contamination of the counter by strongly active samples; if this is large, it may raise the background too high for accurate counting of a weakly active sample following; and it may not always be possible to count the weak sample first. Similarly, decontamination of liquid scintillation counting vials for re-use can be troublesome, and there are many references on the subject (e.g. Drosdowsky and Egoroff, 1966; Harris and Friedman, 1969; Kushinsky and Paul, 1969). Liquid scintillation brings quenching problems, and self-absorption in soft beta emitters can be difficult with end-window G–M counting. Whatever method is used, counting geometry (Section 3.1) must be constant and reproducible, so that results may be uniform and the counts from several standards of known relative activities give the right answers.

4.4.6 Safe handling of isotopes

The need for safe handling is of paramount importance in work with ionising radiations, because of the insidious nature of the danger from them (Section 6.1.2). It can hardly be over-emphasised, and safety must be built into all radiation work right from the start, so that it is an essential factor in planning the experimental design, and the prospective radiation worker must gain a full understanding of safety precautions and the reasons for them before starting work. Safe handling is covered in detail in Section 6.6.

4.5 INTERPRETATION OF MEASUREMENTS

The results of a radiotracer experiment are recorded in terms of count rates (counts per unit time), an arbitrary unit whose meaning depends on the instrumentation being used. In order to be made meaningful, results have to be related to a standard, as in most tracer work, or converted to an expression of the actual amount of radioactivity present. Corrections must therefore be applied to the observed readings, to account for three factors: first, disintegrations in the sample which have not been recorded; secondly, counts arising elsewhere than in the sample; and thirdly, radioactive decay. The recorded count rate is thereby related to the disintegration rate of the sample, i.e. to its activity. It is also necessary to know something of the reliability of the results (not the same as their reproducibility), so that statistical considerations are involved.

4.5.1 Relating the results to a standard

Standardisation is no problem *in vitro*, since an aliquot of the original isotope preparation can be given exactly the same treatment as the experimental samples, and a direct comparison made.

Where the 'sample' is an organ buried in the body of an animal, the comparison becomes more difficult in proportion to the difficulty of reproducing the counting geometry, for standard samples must be counted under the same conditions for a meaningful comparison. The problem is solved by means of a '*phantom*', a full-size model of the organ concerned: constructed (usually) from plastic, for ease of working, it is filled with an aliquot of standard isotope preparation and is either inserted into the body of an inactive animal, replacing the organ it represents, or immersed in water or saline to represent the tissue surrounding the organ. The phantom is then counted under the same conditions as the experimental animal. It is also useful for determining the best arrangement of screening to collimate the radiation from the organ and cut out unwanted radiation from other parts of the body.

4.5.2 Lost counts

The existence of detector resolving time (Section 3.2.3) means that some disintegrations in the sample will not be recorded: this is *coincidence loss*, and its magnitude is easily calculated if the resolving time is known. Suppose that a source would give a count rate N per second if the counter had no resolving time, and that a count rate n per second is actually observed, with a resolving time of t seconds. The counter is dead for t seconds following each count it records, so that it is dead for a total of nt seconds in every second, and it misses Nnt counts every second.

$$\text{Therefore,} \quad N - n = Nnt$$

$$\text{Or,} \quad N = \frac{n}{1 - nt}$$

As mentioned above, resolving time is significant only if it is long and if count rates are high in relation to it. In tracer work, it is normally important only with G–M counting; with a resolving time of 400 μs, coincidence loss is less than 2% at count rates below 3000/min.

Even allowing for resolving time, the detector will not pick up the disintegrations in the sample because it can never be 100% efficient. The various factors affecting counting efficiency, including counting geometry, should best be arranged to be constant, and they may then be ignored if the experimental samples are compared to a standard treated in exactly the same way.

Self-absorption in a soft beta emitter means that the source's 'radiation output', i.e. equivalent or effective activity, is measured, rather than the 'total radiation content' or total activity. The construction and use of a self-absorption correction curve is described in Section 2.4.3.

Backscatter (Section 2.4.3) is another possible variable, predominantly associated with 2π beta counting. It can give a useful increase in count rate of weak sources if they are mounted on a heavy metal support. Planchets tend to vary slightly in thickness, and ideally they should not be thinner than the saturation backscatter thickness for the nuclide being used, otherwise there is a chance that count rates may vary somewhat.

In scintillation counting, Compton scattering of gamma photons degrades them in energy and increases the spread of the photopeak (Section 3.3.3).

Counting efficiency in liquid scintillation counting is greatly affected by quenching (Section 3.3.5), and every counting sample should be corrected for it, by determining the counting efficiency in one of the following three ways.

Internal standardisation. Here the sample is counted before and after addition of a known amount of the same isotope in a non-quenching form; counting efficiency is the difference in count rates compared to the disintegration rate of the added standard. Unfortunately, in order that the composition of the sample shall be affected as little as possible, only small volumes can be added, and they are difficult to measure; also, the method is time-consuming, and the sample and the internal-standard errors reinforce each other.

External standardisation. This is based on the principle that an external gamma source, of known activity, produces Compton electrons in the fluor which are quenched to the same degree as the beta particles from the source being counted. This is the least accurate method of standardisation.

The channels ratio or pulse height shift method. This utilises the fact that quenching shifts the beta pulse height spectrum towards a lower energy level, and the degree of this shift is a measure of quenching. Two channels are selected, one covering almost the whole spectrum and the other about the first third. The ratio of the net sample count rates in the two channels increases with the extent of quenching, as the spectrum shifts, and counting efficiency in a sample is obtained by reference to a standard quench correction curve prepared with samples of known composition. (Dobbs, 1963; Smith and Reed, 1965; DeWachter and Fiers, 1966; Paix, 1966; Rogers and Moran 1966.)

The efficiency of the detector and counting circuitry may show slight variations from day to day, due to slight fluctuations in the voltage supply etc., but these are rarely important. Efficiency can be checked, if desired, by calibration with a standard source of long half-life that gives a fairly high count rate.

4.5.3 Background

Some counts recorded are not due to disintegration of the sample; these background counts arise from cosmic and other natural radiation and from thermionic noise in the circuitry, and are subtracted from the observed total count to give the net source count. This may be the only correction required to an observed count. When background is low and count rates are high, as is often found in G–M counting, it is of no great importance, but otherwise it should always be corrected for. A liquid G–M tube should have a background count taken immediately before every sample count, to correct for any residual contamination of the counter by previous samples. Background is more of a problem with scintillation counting (Section 3.3.3).

Background counts are subject to resolving time equally as are source counts, and the observed result should be corrected for this (if necessary) before subtracting background; it is not strictly valid to do it the other way round, though the error is negligible when source counts are high in relation to background.

Where the ratio of sample counts to background is small, the standard error of the latter becomes important and for best results the total counting time is divided in accordance with the respective count rates, as follows:

$$\frac{\text{Total time}}{\text{Background time}} = \frac{\text{Total count rate}}{\text{Background count rate}}$$

This division applies where the difference in two count rates is required. Where the ratio of two count rates is sought, the formula is:

$$\frac{\text{Time for sample A}}{\text{Time for sample B}} = \frac{\text{Count rate of B}}{\text{Count rate of A}}$$

4.5.4 Radioactive decay during the experiment

This must be corrected for by calculation or by reference to a decay table or graph (Section 2.2.2). *See* p. 278 for worked examples.

4.5.5 Statistical considerations

Because of the random nature of radioactive decay, the accuracy of a measurement depends on the magnitude of the count rate and on the time taken to record it. Recording for a time which is short compared with the average time interval between each event counted will be less accurate than a longer time of recording. A succession of readings n on the same sample, all other things being equal, will give different values of n scattered about a mean, and following a 'normal' or Gaussian distribution. The extent of the scatter of the observations is measured simply by the standard deviation (SD):

$$SD = \sqrt{\left(\frac{\Sigma n^2}{x} - \bar{n}^2\right)}$$

where x = number of values of n, and $\bar{n}$ = mean of all the n's. As the recording time increases, the scatter decreases. With a large number of observations, the mean $\bar{n}$ approaches m, the 'true' value of the count (whose value is rarely, if ever, known; counting a radioactive sample merely estimates it); the SD then simplifies to $\sqrt{m}$ and approximates to $\sqrt{\bar{n}}$. Usually there are one or two observations only, so that a calculation of $\bar{n}$ and of $\sqrt{\bar{n}}$ has little significance. The *standard error* (SE) of a single determination is defined as the square root of the total count, i.e. $\sqrt{n}$, and the *standard proportional error* (SPE) as $\sqrt{n}/n$ or $1/\sqrt{n}$. Thus, as the value of n increases, the SPE diminishes, and can be adjusted to any desired level by collecting enough counts, irrespective of the time taken to accumulate them; the length of time for which a sample is counted depends on this desired accuracy and on the count rate of the sample. A total count of 100 has an SE of 10 and an SPE of 1/10 or 10%; with 1000 counts the SPE reduces to 3·2%, and with 2500 counts to 2%. In biological radiotracer work the 'norm' is usually taken as 10 000 counts, whose SPE is 1%, and this is usually quite adequate. It should be emphasised that there is no point in wasting time collecting counts to reduce the SPE if it is thereby made very much less than all the other errors in the experiment. A common practice by beginners in radiotracer work is to take two or more counts from a sample and average the result—this confers no advantage, and it is just as good, if not better, to spend the same time taking just one count.

Sometimes one feels intuitively that a count is a 'wrong 'un' which does not fit in with the rest. This arises from the fact that the Gaussian distribution has two tails extending a long way from the mean, and observations occasionally fall in the tails: values that deviate from the mean of a series by more than two or three times the *SD* occur with a probability of 4·5% and 0·27% respectively. Retention of such observations could well introduce an unnecessary error into the experiment, and their rejection is in fact justified; a number of more or less elaborate tests, such as the Chauvenet criterion, may be used as the basis for an objective decision.

4.6 OTHER TECHNIQUES

So far, the simple tracer technique has been dealt with, in which a quantity of a single isotope is introduced into an experimental system, and its concentration determined subsequently in samples of the several components of the system. Variations of this simple technique may be employed.

4.6.1 Double (or multiple) labelling

This technique uses two (or more) different isotopes, either of the same or of different elements; they must of course be distinguishable, either by virtue of their radiation properties (half-life, energy and type of radiation), or by one being stable and one radioactive, or by being chemically separable. The term is also used when the same isotope of the same element is introduced into two (or more) different positions in a molecule. Double labelling yields more information than can be obtained with single labelling, i.e. it is possible to study different parts of a molecule simultaneously, or to investigate two-way flows, such as flow of a solute between two compartments separated by a membrane.

Double labelling implies that a molecule is labelled in more than one position, which would be true or molecular double labelling. But, since most preparations consist of a small proportion of labelled molecules and a much larger proportion of unlabelled ones, statistical double labelling is satisfactory and sufficient for most purposes. Here the preparation as a whole shows the desired labelling pattern, though an individual molecule is only singly labelled, if it is labelled at all. Double-labelled preparations may thus be made simply by mixing single-labelled materials, and the same result obtained as though truly double-labelled compounds were being used. A similar consideration applies with U and G compounds (Section 4.1.3): in a sample of glucose-^{14}C (U) the chance of finding a molecule with all six carbons labelled is very slight except at high specific activities. Koch (1968) has suggested a method of simplifying calculations in double-label experiments, and McTaggart and Cardus (1969) assess the errors involved in using five isotopes at once.

4.6.2 Isotope dilution

Isotope dilution is perhaps more a chemical technique, though it does find application in biological work for quantitative analysis. Its principle is that, if a radioactive tracer is mixed with unlabelled compound, its specific activity will be reduced, and comparison of the specific activities initially and finally enables the amount of diluting compound to be calculated. Thus, defining specific activity as total activity divided by weight, the specific activity of the tracer is

$$S_1 = \frac{A_1}{W_1}$$

where A_1 is the activity and W_1 the weight of the tracer. When mixed with a weight of unabelled compound W_u, the specific activity of the mixture will be

$$S_m = \frac{A_1}{W_u + W_1}$$

From this,

$$W_u = \frac{A_1}{S_m} - W_1 \quad \text{or} \quad W_u = \frac{A_1 W_m}{A_m} - W_1$$

where A_m is the activity and W_m the weight of the mixture. If the specific activity of the tracer is high, W_1 will be small and can be ignored, so that the calculation for W_u simplifies to $A_1 W_m / A_m$. The mixing of the tracer and the unlabelled compound must be thorough and complete, or the sample of the mixture that is taken may not be representative.

Purity is important both for the labelled compound and for the separated mixture. Not all of the mixture need be rigorously purified so long as enough of it can be obtained pure to determine the specific activity. If the quantity is too small, it will be difficult to determine by direct weighing, but it can be measured in many cases by spectrophotometry or similar means. Some decomposition of the compound is immaterial, because both the labelled and unlabelled molecules are equally affected and their ratio remains the same.

The purity of the tracer should be as high as possible; it can be checked by reverse isotope dilution analysis. A large amount of inactive pure compound is mixed with the tracer and the mixture is exhaustively purified. The ratio of the specific activities after and before purification indicates the radiochemical purity of the tracer. Calculation of precision in isotope dilution experiments is dealt with by Coulter (1969).

Two variations of dilution analysis may be met with:

1. Derivative dilution analysis—the compound to be determined is quantitatively reacted with a radioactive reagent of known specific activity. The activity of the separated product indicates the amount of radioactive reagent and, since the stoichiometry of the reaction is known, the amount of compound can be calculated.
2. Sub-stoichiometric dilution analysis—for determination of small amounts.

4.6.3 Saturation analysis

Saturation analysis is carried out with a binding agent specific for the compound being determined. An excellent example is the radio-immuno-assay of insulin. A known amount of labelled compound is mixed with that to be determined, and a known amount of binding agent less than that equivalent to the mixture is then added. The inactive and active molecules compete for sites on the binding agent, and the ratio of free to bound activity gives a measure of the unknown compound.

4.6.4 Activation analysis

This is rather different in that the isotope is not put into the experiment at the beginning, but is made at the end, by conversion of an ordinary element to a radioactive form. This is carried out by irradiation of the sample, most commonly with neutrons. A standard containing a known amount of the element is irradiated alongside the sample, and a direct comparison of the two activities gives the amount of element in the sample. The method is most applicable to elements, first, that have a high neutron capture cross-section (Section 2.4.4) relative to the others present, and secondly, which form an isotope whose half-life is longer than the others, for many other elements will be activated besides the one under investigation. Access to a nuclear reactor or other source of neutrons is also essential, though not always easy to arrange, and facilities must be available for handling the 'hot' irradiated samples. The method is not generally applicable to biological materials, but has been used for determination of 'foreign' elements in forensic medicine: 0·3 parts per million of mercury can be detected easily. The most famous use of the method has been the demonstration of comparatively large amounts of arsenic in Napoleon's hair, which is taken as proof that the great man had been steadily poisoned to death.

Activation analysis may also be carried out with gamma radiation or charged particles. Gamma activation analysis (Baker, 1967) is a powerful technique for determination of trace levels of carbon, nitrogen, and oxygen in pure materials; and it may also be used for many other elements. It is a complement to neutron activation, to be used when conditions are favourable, although the characterisation, separation, and measurement of the isotopes produced are likely to be more difficult. A biological application is the determination of iodine; gamma irradiation converts it to iodine-126, which is the only isotope left in sodium chloride media after three days. By contrast, neutron irradiation yields 25-min half-life iodine-128, which has to be chemically separated from the sodium-24 and chlorine-38 formed at the same time.

An application of activation by charged particles was the determination of oxygen in a study of its metabolism by algae. They were allowed to grow in an atmosphere containing stable oxygen-18, and possible metabolites were then separated. The oxygen-18 was then identified and quantified by irradiation with protons, which converted it to radioactive fluorine-18. Chamberlain (1969) suggests that activation analysis of Man may be feasible.

Coleman and Pierce (1967) give a review paper on activation analysis, containing 180 references. Other papers are: Anders and Briden (1964), Bramlitt (1966), Amsel and Samuel (1967), Mulvey *et al.* (1968), Brune (1969), and Malvano and Grosso (1969).

REFERENCES

Books

CHASE, G. D., and RABINOWITZ, J. L., (1962). *Principles of Radio-Isotope Methodology*, Burgess, Minneapolis, 372 pp.
CRAFTS, A. S., and YAMAGUCHI, S., (1964). *Autoradiography of Plant Materials*,

University of California Agricultural Experimental Station Extension Service, Manual 35.

FRANCIS, G. E., MULLIGAN, W., and WORMALL, A., (1959). *Isotopic Tracers*, 2nd edn (o.p.), University of London Press, London, 524 pp.

WANG, C. H., and WILLIS, D. L., (1965). *Radiotracer Methodology in Biological Science*, Prentice-Hall, New York, 382 pp.

The Radiochemical Manual, (1966). 2nd edn, The Radiochemical Centre, Amersham, Bucks., 327 pp.

A somewhat specialised text is:

Advances in Tracer Methodology, **3**; *Proc. 9th 10th Symp. Tracer Methodology*, (1966). Ed. ROTHCHILD, S., Plenum, New York, 333 pp.

Papers

AMSEL, G., and SAMUEL, D., (1967). 'Microanalysis of Stable Isotopes of Oxygen by Nuclear Reactions', *Analyt. Chem.*, **39**, 1689–1697.

ANDERS, O. U., and BRIDEN, D. W., (1964). 'Rapid Non-Destructive Method of Precision Oxygen Determination by Neutron Activation', *Analyt. Chem.*, **36**, 287–292.

BAKER, C. A., (1967). 'Gamma-Activation Analysis', *Analyst, Lond.*, **92**, 601–610.

BRAMLITT, E. T., (1966). 'Chlorine Analysis by Neutron Activation', *Analyt. Chem.*, **38**, 1669–1674.

BRUNE, D., (1969). 'Epithermal Neutron Activation for Iodine in Small Aqueous Samples', *Analyt. chim. Acta*, **46**, 17–22.

CHAMBERLAIN, M. J., (1969). 'Activation Analysis of Man', *New Scient.*, **43**, 575–579.

COLEMAN, R. F., and PIERCE, T. B., (1967). 'Review Paper—Activation Analysis', *Analyst, Lond.*, **92**, 1–19.

COULTER, B. S., (1969). 'Calculation of Precision in Isotope Dilution Experiments', *Int. J. appl. Radiat. Isotopes*, **20**, 271–274.

DeWACHTER, R., and FIERS, W., (1966). 'External Standardisation in Homogeneous Liquid Scintillation Counting of Tritium, Carbon-14, and Phosphorus-32' *Archs int. Physiol. Biochim.*, **74**, 915–916.

DOBBS, H. E., (1963). 'Measurement of Liquid Scintillation Counting Efficiency' *Nature, Lond.*, **200**, 1283–1284.

DROSDOWSKY, M., and EGOROFF, N., (1966). 'Apparatus for Decontamination of Scintillation Counting Vials', *Analyt. Biochem.*, **17**, 365–368.

GERRARD, M., (1969). 'Tagging of Small Animals with Radio-Isotopes for Tracking Purposes. (A Review)', *Int. J. appl. Radiat. Isotopes*, **20**, 671–676.

HARRIS, J. E., and FRIEDMAN, L., (1969). 'Rapid Inexpensive Decontamination of Scintillation Counting Vials', *Analyt. Biochem.*, **30**, 199–202.

KOCH, A. L., (1968). 'Method to Simplify Calculations in Double Label Experiments', *Analyt. Biochem*, **23**, 352–354.

KUSHINSKY, S., and PAUL, W., (1969). 'Cleaning of Liquid Scintillation Vials', *Analyt. Biochem*, **30**, 403–412.

MCTAGGART, W. G., and CARDUS, D., (1969). 'Analysis of Errors in a Technique for Combined Use of Multiple Radio-Isotopes', *Int. J. appl. Radiat. Isotopes*, **20**, 420–436.

MALVANO, R., and GROSSO, P., (1969). 'Iodine Activation Analysis in Materials of Biological Interest', *J. nucl. Biol. Med.*, **12**, 86.

MULVEY, P. F., CARDARELLI, J. A., ZOUKIS, M., COOPER, R. D., and BURROWS, B. A., (1966). 'Sensitivity of Bremsstrahlung Activation Analysis for Iodine Determination., *J. nucl. Med.*, **7**, 603–611.

PAIX, D., (1966). 'Do-it-Yourself External Standardisation for Liquid Scintillation Counting', *Int. J. appl. Radiat. Isotopes*, **17**, 486–487.

RESCIGNO, A., (1968). 'Flow Graphs and Tracer Kinetics', *J. nucl. Biol. Med.*, **12**, 59–67.

ROGERS, A. W., and MORAN, J. F., (1966). 'Evaluation of Quench Corrections in Liquid Scintillation Counting', *Analyt. Biochem.*, **16**, 206–219.

SIMON, H., and PALM, D., (1966). 'Isotope Effects in Organic Chemistry and Biochemistry', *Angew. Chem.—int. Ed. Engl.*, **5**, 920–932.

SMITH, A. H., and READ, G. W., (1965). 'The Standardisation of a Liquid Scintillation System', *Proceedings of the Symposium on Radio-Isotope Sample Measurement Techniques in Biology and Medicine*, International Atomic Energy Agency, Vienna, 427–445.

SMITH, C. U. M., (1968). 'Discrimination Between Heavy Water and Ordinary Water by the Mouse', *Nature, Lond.*, **217**, 760.

Chapter 5

THE USES OF IONISING RADIATION

The principles of the isotopic tracer technique have been dealt with in the preceding chapter, and this one will deal with some of the other uses and applications of ionising radiations that are of interest to the life scientist. A large part is taken up with ways in which the damaging effect of ionising radiation on living organisms and tissues may be turned to good advantage: sterilisation of equipment, prevention of food spoilage, pest control, and radiotherapy.

5.1 STERILISATION OF EQUIPMENT (IAEA, 1967–1)

Surgical equipment is traditionally sterilised by heating, either dry (baking), or wet (steaming, autoclaving), or by immersion in alcohol. If the equipment is used more than once, it has to be cleaned and re-sterilised every time—the costs involved can, however, be obviated with disposable equipment which is used once and thrown away; this also cuts out any possibility of cross-contamination. Mass-produced plastics items are cheap enough to make disposable equipment economical, but there is the problem of ensuring sterility, for many plastics melt on heating and/or soften or dissolve on immersion in alcohol, and they cannot therefore be sterilised by these methods. Gaseous ethylene oxide is usable with plastics, but has the disadvantage that it only sterilises surfaces, and any dirt on the equipment will have a nicely sterile surface concealing a contaminated area below. Radiation overcomes the difficulty, for it is penetrating, and plastics having the correct proportion of cross-linked long-chain molecules in their structure are not harmed by it. It is a very reliable and efficient means of sterilisation, doses of $2{\cdot}5\times10^6$ rad (the rad is defined in Section 6.2) ensuring that even the most resistant bacterial spores have a chance of survival of less than one in 10^{15}. Radiation sterilisation provides another advantage, for although plastics items are sterile when they emerge from the moulding machine (having been formed by heat treatment), maintenance of sterility cannot be guaranteed between then and sealing into their individual packages. Sterilisation after packaging overcomes the problem and, once sterile, the items will remain so until the package is opened at the time it is required for use. Radiation-sterilised syringes, needles, etc., are now used by both large and small consumers, hospitals, general practitioners, laboratories, and others.

The gamma radiation from cobalt-60 or caesium-137 is used in sterilisation plants; these isotopes are now available comparatively cheaply as by-products of the uranium fuel from nuclear reactors. Sources of 5×10^4–5×10^5 Ci are employed, heavily shielded by some 5 ft of concrete, and operation is almost entirely automatic and remotely controlled. High-speed electrons, produced by electrical machines, may also be used for sterilisation, but they are more suited to the treatment of sheets of material, since their penetrating power is not great.

5.2 PREVENTION OF FOOD SPOILAGE

Stored food is very easily spoiled, for it is attractive to many other organisms besides Man, and their presence and activities then render the food unfit for human consumption. A disturbingly large proportion of the world's food is wasted in this way, and it is good that the harmful effects of ionising radiations on living things can be put to use in preventing food spoilage. This application of radiation may be considered under five headings as follows:

1. Complete sterilisation, the total destruction of all bacteria, micro-organisms, and spores, for long-term storage;
2. Pasteurisation, or partial sterilisation, when vegetative forms of organisms are inactivated, although spores may still remain;
3. Disinfection, the inactivation of certain pathogenic micro-organisms;
4. Disinfestation, by sexual sterilisation of larger organisms (e.g. insects);
5. Inhibition of sprouting of stored root crops.

The cost of preservation of foodstuffs by irradiation must be considered in relation to the other available methods. The capital cost is high, although running costs are low, so that radiation preservation is only applicable where large quantities are involved, and the initial capital outlay can be spread out sufficiently to avoid a prohibitive price increase to the consumer (Josephson, Brynjolfsson, and Wierbicki, 1968).

Sterilisation. The complete destruction of bacteria and micro-organisms in food by radiation treatment was promising at first, but it was soon found to have limitations. The killing dose for bacteria causing food spoilage is generally quite low, but a bacterium of high radioresistance must be taken as an index of the efficiency of processing. The most resistant that can be found in food is *Clostridium botulinum*, which also happens to be one of the most dangerous bacteria known for, while it does not make the food uneatable, it produces highly toxic metabolites that cause the type of food-poisoning known as botulism. Enzymes responsible for long-term food spoilage need even higher doses for inactivation, but they can be inactivated equally well by heat, so that cooking in the conventional way would seem to be as good a way as any for preservation of food by sterilisation. Radiation in extremely high doses will certainly sterilise food, but at the same time it will render the food unacceptable for human consumption in various ways which are dealt with below. On the other hand, the sterilisation of food wrappings and packages is quite acceptable, since they are not eaten and can therefore have a full sterilising dose of radiation without any ill effects being of significance to the consumer of the food. It may be noted that prior heat treatment reduces the radioresistance of many organisms; radioresistance is also related to the water in the food.

Effects of ionising radiation on food

Induced activity. Besides affecting living organisms in food, irradiation may also affect the food itself by inducing radioactivity. Heavy particles (protons, deuterons, alphas, neutrons, etc.) cannot be used as the activity they induce is long-lived; gamma radiation can also induce activity if it is above a certain threshold energy. The products from gamma irradiation of most elements commonly found in foods have short half-lives, except iodine-126 (13 d) from iodine-127, so that, apart from this, there is little problem with gamma irradiation. For foods containing iodine, a minimum time of storage between irradiation and consumption must be specified, to allow the iodine-126 to decay away. Induced activity from electron irradiation has been shown to be well below the maximum concentrations permitted in food, except with carbon and chlorine;

the half-lives here are short (20 and 33 min) so that they present no danger.

Heating effect of radiation. 10^6 rad is equivalent to about 2·4 cal/g (1 cal=4·2 J) of the irradiated material, the rise in temperature being inversely proportional to the specific heat. There is thus a possibility that certain foods of low specific heat may be damaged by heating effects, if given large doses of radiation without any means of cooling at a relatively high initial temperature.

Biochemical effects. It may be concluded that the biological or nutritive value and the digestibility of foods are unaffected by radiation, or slightly lowered to an extent comparable with that given by other accepted treatment methods, although extended long-term studies would perhaps be useful. Difficulty arises through the production of undesirable by-products of radiation action that make the food unacceptable.

Most amino acids are affected by high radiation doses: several, including leucine, valine, lysine and arginine, are thought to be a possible source of objectionable odours and flavours after irradiation. Leucine yields isovaleraldehyde, which has a strong 'goaty' odour, and tryptophan yields indole or related compounds; these and the sulphur compounds from irradiation of methionine and cystine also smell unpleasant. A general reaction of carbohydrates is oxidation of alcoholic to carbonyl groups, thought to be responsible, for example, for the development of brown colour in milk. Polysaccharides are depolymerised and oxidised, leading to softening of the food. Fats undergo peroxidation

$$\text{—COOH} \rightarrow \text{—}\underset{\underset{\text{O}}{\|}}{\text{C}}\text{—O—OH}$$

and carbonyl group formation, more so in the presence of oxygen than in its absence. Ingestion of carbonyl compounds has been thought to produce toxic effects, but no gross effects were noted in rats after a month on a diet very high in such compounds. Vitamins all seem susceptible to breakdown by radiation, particularly vitamins C and E, their sensitivity being greater in the pure state than when incorporated into food. There are many conflicting results and little reliable information, so that any general statements about the vitamin content of irradiated foods are difficult or impossible to make. Much more research needs to be done to clarify the picture.

Organoleptic changes. Radiation often induces undesirable changes in taste and/or appearance of foodstuffs, which render some foods totally unsuitable for radiation treatment.

With meat, the limiting factor for treatment is thought to be the development of rancidity in its fats, and the development of bitter, metallic, or burnt flavours: it seems unlikely that large cuts of meat can be effectively treated.

Fish in general appear to be less sensitive to change than meats. The shelf-life under refrigeration of all types of fish, including shell-fish, can be extended by up to 1 or 2 months by doses of 5×10^5 rad; this is pasteurisation rather than sterilisation, the organisms in the food being prevented from reproducing rather than being killed outright. Higher doses cause burnt flavours and discoloration with fish.

Fruits are very prone to spoilage by radiation: there are many conflicting results, so it is difficult to draw useful conclusions, but softening (from the effect on the polysaccharides) is perhaps the most difficult problem. Most promising results are obtained with apple juice and strawberries: 5×10^5 rad will destroy mould on the latter and double their storage-life, which is quite an achievement for such delicate fruit. Surprisingly little work has been done on vegetables.

Milk gives very disappointing results, quickly developing unpleasant taste, odour and colour. Eggs have been said to give quite promising results, but it must be remembered that the problem in prolonging storage-life here is not to kill organisms, for the interior of a new-laid egg is sterile and there are none to kill, but to prevent them getting in. Methods of sealing the shell would therefore seem to be more applicable than those for sterilising the interior.

Other effects. Toxicity tests on irradiated foods have proved negative, so that they can be assumed not to present a toxic hazard (Kraybill and Whitehair, 1967). Much work has been done on their possible carcinogenicity, but the only general conclusion to have emerged is that each food must be given a thorough feeding test and a minute analysis, and the results must not be extrapolated to other foods. It has been reported that genetic changes have occurred in *Drosophila* which have been fed irradiated food, but the extrapolation of fruit-fly results to man is of very dubious validity (Rinehart and Ratty, 1967). Schubert (1969) considers that the usual feeding tests for irradiated foods may not be adequate and that supplementary tests are desirable.

Pasteurisation

With food, pasteurisation shows more promise than complete sterilisation, since it requires smaller doses and preservation is accomplished without the radiation adversely affecting the food itself.

The same is true of disinfestation and to some extent of disinfection.

Disinfection

Disinfection is applicable in so far as the very common causative organisms of food poisoning, *Salmonella*, are concerned; they are the most sensitive of the pathogenic organisms in food, and doses of 0·5–0·7 Mrad are adequate for complete inactivation. Viruses, though, are rather more resistant.

Disinfestation

In meat, control of tapeworms and trichinosis requires low doses only, and is effective and economical, especially where incidence of these parasites is high.

Stored cereals can be greatly damaged by flour and grain beetles and weevils, but only if the pests are present in large numbers; a few will not cause significant loss of the food. If these few can be prevented from reproducing, the object of the operation will have been achieved and, whereas it takes 10^6 rad to kill them outright, only 2×10^4 rad are needed to sterilise them and to prevent any eggs and larvae that may be present from maturing. The only noticeable effect on the grain of 2×10^4 rad is that a little more water may be needed to make dough, arising from a slight alteration in the physical properties of the starch (Cornwell, 1966).

Stored root crops are liable to spoilage, not only from attack by other organisms, but also by their own metabolic activity which leads to sprouting. A dose of 10^4 rad will prevent potatoes sprouting, and lengthen their storage-life considerably; the cooking time may be slightly shortened because of the tenderising effect of the radiation on the cellulose. Carrots and onions have also been investigated.

Summarising, it may be concluded that, except in a few cases, the overall picture of the possible harmful effects and the feasibility of treating food with radiation is still far from clear; and it is accepted that, until it is, the only course of action is a total ban or radiation treatment, except as specified foods irradiated under specified conditions are proved without doubt to be satisfactory for human consumption. Each food submitted for clearance and exemption must have all the factors in its treatment specified in great detail: the energy, type, dose and conditions of irradiation; the nature of the containers in which it is packed; the conditions of storage, minimum time that must elapse between irradiation and consumption, and maximum store life; and the effect of irradiation on nutrients, especially vitamins and essential amino acids. Once these details have been laid down, no deviation from them is permitted. Tests for toxicity, carcinogenicity, induced radioactivity, and the microbiological hazard must have been carried out and shown to be satisfactory.

The United States of America has released sterilised bacon and disinfested wheat for human consumption, and Canada allows the disinfestation of wheat and the inhibition of sprouting in potatoes. In the United Kingdom, irradiation of food is prohibited except for patients who require a sterile diet as an essential factor in their treatment (S.I., 1969). It is likely that the use of ionising radiation in food preservation will in time prove very useful, but more as a supplement to than a replacement for any of the more well-established methods (FAO/IAEA, 1966; Reber, Raheja, and Davis, 1966; IAEA, 1967–2; Kraybill and Whitehair, 1967; Bowman, 1968).

5.3 PEST CONTROL

The use of ionising radiation for pest control makes use of the fact that a comparatively small dose of radiation will render the pests unable to produce viable gametes. They will be able to mate, but the union will be infertile and no offspring will be produced. The application of the principle was seen in the previous section, in the disinfestation of stored grain from beetles and mites.

It is seen also in the so-called *sterile male* technique for control of insect pests in the field. A very large number (several millions) of male insects are bred in the laboratory, and are sterilised by a suitable dose of radiation. They are then released into the area where control is desired, where they mate with the indigenous females who then produce no offspring. Success in using the method depends on the fulfilment of several conditions. The number of sterile males released must be comparable with the total population already present, so that the chances of a female having a sterile mating shall be reasonably high. The females should mate only once in their lifetime; fortunately many insects do just this. The insects should not be migratory for, while it may be easy to clear an area of the pest, it will be in vain to do so if the area is repopulated by normal insects moving in from neighbouring regions. However, there are several instances where these conditions have been successfully satisfied: pests have been brought under control and, by repetition of the technique as necessary, their numbers have been reduced almost to vanishing point (e.g. Abdel-Malek, Tantawy, and Wakid, 1966; IAEA, 1968).

5.4 RADIOTHERAPY

The best means of dealing with a cancerous growth in the body is undoubtedly its complete surgical removal. Ionising radiation present a possible alternative in cases where surgery is undesirable or not possible, and this brings home the ironic unfortunate truth that the same radiations that cure cancer may also cause it (Section 7.4.4); like most swords, radiations have two edges.

5.4.1 Principle

In the same way that control of crop pests by irradiation is achieved by preventing them from reproducing, so in radiotherapy the cancerous cells must be prevented from proliferating indefinitely, so that their malignant uncontrolled growth is stopped, even if they divide once or twice after irradiation before they die (Section 7.3.5); this done, the patient is adjudged cured as effectively as if the cancer had been surgically removed, even though the symptom, i.e. the existing growth, is still present. Some growths are difficult to deal with, for a single cell left alive may be able to reproduce and cause a recurrence of the growth; all cells in such a tumour must be truly killed to achieve an effective cure.

Unlike the surgeon's scalpel, radiation is not fully selective in its effects on living tissue, for it will damage healthy and cancerous tissues alike. The limiting factor in radiotherapy is the dose of radiation which the healthy tissue can withstand rather than the dose required to kill the cancer, and the problem is to minimise the damage to the healthy tissue while maximising that to the cancer. This problem is most easily solved if the source of radiation can be placed as close as possible to the cancer: the inverse-square law operates, so that the radiation dose is very intense locally and falls off rapidly as the distance from the source increases. If a suitable source is chosen, the cancer can be lethally irradiated while the radiation dose to the healthy tissue surrounding it is negligible. Contact therapy, interstitial or intracavitary irradiation, and intravenous treatment have been developed on this principle.

5.4.2 Contact, interstitial (intracavitary), and intravenous therapies

Contact therapy is applied to the superficial areas of the body, including the eye and such cavities as the mouth, nose, throat, etc. The isotopes are sealed into suitably shaped applicators, depending on the area to be treated. They are specified in terms of their activity, the dose rate they give, and their 'radium equivalent'. Pure beta emitters used are phosphorus-32, whose fairly hard radiation penetrates to a depth of about 3 mm in tissue, and yttrium-90, either alone or as its long-lived parent strontium-90. Gamma radiation penetrates to a greater depth, the isotope used being cobalt-60. Its beta radiation is screened out by a platinum sheath.

Interstitial (intracavitary) therapy is used for regions within the body. The sources may be implanted and withdrawn after treatment for a suitable length of time; they are usually sheathed in inactive material such as platinum, and also specified in terms of their 'radium equivalent'. Isotopes used are cobalt-60, caesium-137, tantalum-182, and iridium-192, in a variety of shapes and sizes. Before the advent of the artificial isotopes, radium-226 and 'seeds' containing radon gas were used and, though radium is still available, radon has been entirely superseded, mainly on the grounds of safety. Gold-198 is used for areas that are accessible for only a short time, from which sources cannot be withdrawn: its short half-life, $2\frac{1}{2}$ d, means that it can be left *in situ* after irradiation without harming the patient. An alternative approach is the infiltration of a colloidal solution of isotope into the tumour by direct injection: here, the short half-life isotopes gold-198, phosphorus-32, and yttrium-90 are particularly suitable.

At one time it was hoped that it might be possible to administer an isotopic preparation, either intravenously or orally, that would be selectively taken up by a cancerous tissue, which it would then destroy. These hopes have not been fulfilled, for no substance has yet been found that concentrates specifically in cancers. This is largely because cancer cells are very similar metabolically to the normal cells from which they are derived, the only difference being essentially that, whereas normal cells divide either not at all or in a controlled manner, the division of cancer cells is uncontrolled. However, the approach has had limited success in specific cases, but is restricted by the fact that normal healthy tissue is affected as well as the cancer. The best results have been obtained with the use of radio-iodine to cure disorders of the thyroid gland, which concentrates the element far more than any other part of the body. The destruction of healthy tissue is here of less importance than in most other organs, for it is easy to maintain a patient indefinitely on replacement therapy with suitable doses of thyroid hormones.

A quinol compound, known as 'Synkavit', is obtainable labelled with tritium to a specific activity of more than 30 Ci/mmol, and it appears to concentrate somewhat in tumour tissue. Whether this differential concentration can be exploited sufficiently to kill cancerous tissue without damage to healthy tissue remains to be seen. Boron has been found to accumulate in brain tumours, but not in normal brain cells, and neutron irradiation then converts it to lithium with the expulsion of an alpha particle: the tumour is thus subjected to alpha irradiation. Slow neutrons must be used, or the whole of the brain may be irradiated with protons (Section 2.4.4).

A further possibility is to make antibodies specific to tumour tissues and then to incorporate a suitable isotope into the antibody. Here again success is dependent on the cancerous and the normal cells being sufficiently different biochemically to form different antibodies.

5.4.3 Teletherapy

In this, the other method of radiotherapy, the source is at a distance from the area to be treated. Localisation of the radiation in this area is more difficult than with contact or interstitial therapy, since the inverse-square law is much less important. The radiation beam must be carefully collimated, and healthy tissues must be shielded. The problem is especially acute with deep-seated tumours, for the intensity of radiation from an external source is considerably reduced through absorption by the intervening tissue before it reaches the tumour and, in giving a damaging dose to the tumour, the overlying healthy tissue would be damaged far more. Radiobiological theory indicates several possible approaches which may be able to help in solving the problem, though their translation into practical terms must still be left to experience and empirical observation.

In general, cells which are dividing and which are undifferentiated (unspecialised) are more sensitive to radiation than those which are not. This is fortunate for radiotherapy, since cancers are dividing and are also relatively unspecialised, so that irradiation brings about some preferential destruction of cancerous tissue. The aim of the radiotherapist is to maximise the differential, by suitable choice of energy and dose of radiation, etc.

The *oxygen effect* may be useful (Section 7.5.2). Many tumours are poorly vascularised and are therefore anoxic so that, if their blood supply can be increased, they will receive more oxygen and their radiosensitivity will increase. The blood supply of surrounding tissue will be increased at the same time, but it is likely to be sufficiently oxygenated already for its radiosensitivity to be at a maximum, so that it will not be increased further by increasing the oxygen supply. (Vaeth, 1968; and Vandenbrenk, 1968.) However, it is still not altogether clear if the oxygen effect is pertinent to clinical radiotherapy (Alper, 1968).

Fractionation of the dose has been found empirically to give a better differential between tumour and healthy tissue. A higher total dose is needed, since the tissue can effect some repair of the radiation damage between the fractions, but it has been claimed that the tumour cells have less ability to repair damage than the normal ones.

The so-called '*cross-fire*' or *rotation* technique is valuable with deep-seated tumours. Irradiation is carried out from several different directions, the source being rotated round the patient (or vice versa), the irradiation in each position being no greater than the maximum which the healthy tissue can tolerate. The cancer is arranged to be at the centre of rotation, and simple geometry will show at once how it receives far more radiation than the healthy tissue.

The physical properties of the radiation affect the dose received by the tissue. The energy released at a point in the irradiated volume depends not only on the energy of the radiation, and on the direct radiation reaching it, but also on the scattered radiation, this including both scattered primary radiation (x-ray or gamma photons are used), and the secondary electrons produced by its interaction with the tissue. A wide primary beam is scattered more than a narrow one, so the radiation at a point on the axis of a beam increases with its area. Due to scattering and interaction, the dose at the surface of an absorber is higher than the dose in air would be at the same distance from the source (the 'in-air' dose), though, with low-energy radiation, the surface dose is not greatly increased over the in-air dose. Up to 200 keV energy, backscattering of the photons is important, with a corresponding increase in the surface dose; above this energy, scattering and the emission of secondary electrons tend more and more to be forward, in the direction of the primary beam. This has the result that, above 1 MeV energy, the dose at the surface is not the highest received by the tissue: with 2 MeV radiation, the maximum is three times higher than the surface dose and is received at a depth of 4 mm and, with 22 MeV, the peak dose depth is at about 40 mm. This is of some practical importance, for it is possible to decrease the relative dose to the healthy tissue by selecting the energy of the radiation so that its peak energy release coincides with the depth at which the cancer is situated. The dose at other depths will still be appreciable, so that healthy tissue is not completely spared from radiation. Tables and 'isodose charts', in which points receiving equal doses are linked by lines (exactly as contour map or rainfall charts) can be experimentally determined for particular conditions of irradiation, to show the depth–dose relationship, i.e. how the dose received varies with position in the tissue. They only apply to the conditions under which they were made, so that they must specify peak energy of the radiation, its half-value thickness, the area of the beam, and the distance of target from source (TSD).

The *sources* used for teletherapy are gamma emitters with fairly long half-lives, or x-ray machines. Classically, radium-226 was the isotope used, but caesium-137 and cobalt-60 are used today as they are available cheaply in comparatively large quantities as by-products from nuclear reactors. Sources of 20–2000 Ci are used. Cobalt-60 is supplied at a specific activity of up to 150 Ci/g; its 1·17 and 1·33 MeV gamma energies give a maximum dose a few millimetres below the surface of the skin. Caesium-137 is supplied as the chloride, and has a lower specific activity, 25 Ci/g. Its lower gamma energy, 0·66 MeV, gives more skin injury than cobalt, but its half-life is 30 years compared with only 5·26 years for the former.

An x-ray set is more difficult to use than an isotope, but has the distinct advantage that it can be switched off after use, whereas an isotope is an ever-present hazard. Also, the x-ray energy spectrum can be varied to suit the requirements of each case (Section 2.5.1), whereas isotopic gamma radiation is mono-energetic or restricted to a few well-defined energies.

Deeley and Wood (1967), and Andrews (1968) are recent textbooks on radiotherapy, also Meredith (1968).

5.4.4 Other radiations

Fast neutrons are able to induce highly ionising radiation deep within the irradiated material (Section 2.4.4), and they have therefore been suggested as therapeutic agents (Entzian *et al.*, 1966; Broerse and Barendsen, 1967; Field, Jones, and Thomlinson, 1968). Ultraviolet irradiation of internal organs *in vivo* is possible by application of an external radiation source, e.g. x-rays interacting with calcium fluoride inserted temporarily into rat intestine (Rusznyak, 1968).

A new type of accelerator constructed recently can produce, thousands of times more intensely than ever before, a particle whose properties hold great promise for radiotherapy of tumours

(Rosen, 1969). This particle is the negative *π-meson* or negative *pion*, which carries one elementary negative charge and has about 300 times the mass of an electron. Like all charged particles, pions have an almost unique range for a given energy, and a substantial part of their energy is effective at the point where they stop. Since the shape of their spectrum can be accurately tailored, energy can be deposited just where it is needed, unlike x-rays: in addition, there is no 'exit dose' from the beam as it leaves the tissue. The pions are slowed down by ionisation and, as they stop, they are each captured by the nucleus of a heavy atom in the tissue (i.e. an atom other than hydrogen). As a result of the strong forces involved, the pion mass is converted into energy, which violently disrupts the nucleus; a substantial part of this energy appears as kinetic energy of protons and heavier ions, which travel a very short distance and have a devastating effect on the cells they pass through. Thus with negative pions the tumour dose can be maximised and the dose to the healthy tissue can be minimised. The ratio of tumour to healthy tissue damage is about 10 times that for 8 MeV x-rays and cobalt-60 gamma rays, and about 3 times that for alpha particles.

5.5 RADIOGRAPHY AND RADIO-PHARMACEUTICALS

Whereas radiotherapy is the treatment of disorders by means of ionising radiations, radiography refers to their diagnosis. It shares the feature that it may be carried out by sources either within or without the body, and either isotopes or x-ray machines may be used. The amounts of radiation required are two orders of magnitude less than for therapy. Accurate dosimetry is important in diagnostic radiology, both to get a good picture and to minimise the dose to the patient (e.g. Ardran and Crooks, 1968). The dose delivered may be reduced considerably by the use of an *intensifying screen* placed adjacent to the film on which the image is recorded; this contains a substance, e.g. calcium tungstate, that fluoresces under x-rays, and the fluorescence has a greater effect on the film than the x-rays themselves. An example of radiography is the mass chest x-ray used to screen the general public for early detection of tuberculosis; x-rays are also in common use in dentistry. Marchal *et al.* (1966) describe a radiographic method for instantaneous and continuous recording of the movement of thoracic organs.

There are some disadvantages in using an x-ray machine, for it may be awkward or impossible to reach some positions with it, and also it cannot be made easily portable: a 'flying doctor' working in an isolated area could not transport such a large heavy machine round with him and, even if he could, he would probably have to transport its power supply as well. However, isotopes can come to the rescue: sources for gamma radiography are available, containing a few curies of cobalt-60, caesium-137, iridium-192, or thulium-170. Smaller sources have been developed for dental radiography, of iodine-125 or thulium-170: they are held in the mouth, and enable a radiograph of all the teeth to be taken at one time, which is not possible with x-rays. It may be noted that thulium is a beta–gamma emitter, and it is the bremsstrahlung from the beta, rather than the gamma, which enables it to be used as a radiography source.

The radioactive preparations used for diagnosis internally are commonly known as *radio-pharmaceuticals*. Their use falls into four broad groups, as follows:

1. Determination of body composition, e.g. body water, using dilution analysis;
2. Physical tracing of the circulation, e.g. to see if there is a blockage;
3. Isotopic tracing, e.g. of the absorption and metabolism of biochemicals;
4. Scanning an organ, using a reagent that is picked up by the functional part of an organ but not by a non-functional part.

Many considerations influence the choice of isotope and its chemical form, though it should emit soft gamma radiation so that it can be detected externally but give a low radiation dose to the patient—hard beta and hard gamma should be avoided. Iodine-125 (Haymond, Haig, and Kimball, 1966), mercury-197, and technetium-99m have been successfully used. The isotope should also be excreted rapidly and completely when the diagnosis is complete: the best example here is xenon-133 (used in physical tracing) which is removed from the blood by a single passage through the lungs. A difficulty arises from the fact that most biologically important compounds contain only the six elements C, H, N, O, P, and S, and these have no convenient gamma-emitting isotopes, so that isotopic labelling is possible only in a few cases, and usually foreign labelling (Section 4.4.3) has to be resorted to: e.g. iodination of proteins. Selenium-75 offers possibilities in labelling sulphur compounds, as the two elements behave similarly, but formidable synthetic problems would have to be overcome.

Purity is important with all pharmaceuticals (as indeed it is also with therapeutic agents), from two viewpoints, that of the job which they have to do, and with regard to the safety of the patient. Because they are radioactive, radio-pharmaceuticals present more difficulty than ordinary ones in purity control and, the shorter the half-life, the more acute the difficulty becomes. Testing a batch of product for sterility and freedom from pyrogens takes time and, by the time the tests are complete, the activity may well be too low to carry out the diagnosis. Testing for purity may therefore have to be done retrospectively, along with stringent precautions in manufacture and (presumably) the hope that nothing goes wrong on the way.

5.6 AGE DETERMINATION, OR DATING

Naturally occurring radio-isotopes provide an elegant and precise method, commonly called dating, for determining the age of the materials in which they are found. It is essentially a tracer technique, but it uses tracers introduced by the hand of Nature. The activity in the material is measured and, knowing the half-life and other necessary data, it is converted into the age of the specimen. It must be clearly understood, though, that the age is not measured directly—a set of assumptions are made and used as the basis for calculating age from measured radioactivity.

Any radioactive element occurring naturally may be used for age determination, though it goes without saying that a suitable element must be present in the material to be dated, apart from any other considerations. Elements of very long half-life are applicable to age determinations on a correspondingly long time-scale, and find a biological application in palaeontology; a shorter time scale, using isotopes of shorter half-life, is applicable in anthropology.

Use of isotopes of very long half-life

When using isotopes of very long half-life, the principle is to determine the amount of disintegration products that are present; the time taken to form this quantity is then calculated, and taken as the age of the specimen. It is assumed that: first, all the product elements in the material were formed solely by decay of the parent; secondly, none were present initially; and thirdly, none have been lost from the sample.

Elements having a very long half-life are uranium and thorium, whose decay series end up with lead and which generate helium on the way. The amounts of stable lead-206 or lead-207 (from uranium) or lead-208 (from thorium), or of helium, may be measured; alternatively, since uranium-235 and -238 decay at different rates, lead-207 and -206 will be formed at different rates, and their ratio gives a measure of the time taken to form them. Rubidium-87, which makes up more than a quarter of the naturally occurring element and decays to strontium, and potassium-40, which decays to argon, may also be used for age determination.

A novel approach described by Huang and Walker (1967), is based on the presence of traces of uranium and thorium at about one part per million in many minerals. Tracks are formed in mica and other crystals by the recoil of a nucleus as it emits an alpha particle; the tracks increase in number as the specimen ages and, though very small, they are visible under phase microscopy after etching with hydrofluoric acid. The method is analogous to the previous 'fission-track' method used in the age range $20–1{\cdot}5\times10^9$ years, but is about three orders of magnitude more sensitive. It is suggested that it could be applied to the determination of the age of apatite crystals in fossilised teeth.

Use of isotopes of shorter half-lives

The naturally occurring isotopes of shorter half-lives, formed by the induction of activity in stable elements by the action of cosmic radiation, may also be used for age determination. Here, the quantity of the nuclide itself, rather than of its breakdown products, is determined. Among other things, it is assumed that the cosmic radiation flux, and hence the rate of formation of the nuclide, is constant. The isotopes that may be used include chlorine-36, e.g. to determine the time for which rocks have been exposed on the surface of the earth, and tritium; but perhaps the best-known and most widely used method is *radiocarbon dating*, with carbon-14; its development is chiefly associated with the name of Libby (1955), who was awarded a Nobel Prize for his work.

All atmospheric carbon dioxide is radioactive, and all living things are in equilibrium with it. At death, the organism is cut off from its supply of CO_2, and interchange ceases, so that the activity of the carbon trapped inside the dead body of the organism begins to fall. Determination of the specific activity of a once-living specimen and its comparison with that of atmospheric CO_2 allows calculation of the time since it died. It is assumed, of course, that atmospheric CO_2 has the same specific activity now as it had then—there is, however, evidence to show that this is not true. Several archeological discoveries in ancient Egypt were dated by the carbon-14 technique, and the dates were consistently a few hundred years in error in comparison with the known ages, which had been determined incontrovertibly by other methods. The only possible conclusion was that the activity of carbon-14 in the atmosphere had changed over the past few thousand years. It must also be noted that the method can only date the time at which the material ceased to be living, and not the date at which it was in use, e.g. dead heart-wood may remain within a tree for a century before being used. Radiocarbon dating has been applied principally to the objects found in early human settlements; it was recently extended to collagen residues in bone (Barker, 1967); and it has also been applied to the determination of flood and ocean levels: if a fertile area is flooded, it is cut off from its supply of carbon-14, and the specific activity of its organic matter will decline in the same way as that of a once-living organism (Berger and Libby, 1969). The method is most useful in the age range 600–10 000 years. It presents some technical difficulty in that the radiation is pure soft beta (E_{max} 0·158 MeV), and the specific activity, even of atmospheric or 'free' CO_2, is low, of the order of some 15 disintegrations per min per g of carbon. The activity is therefore measured by a method giving the best possible efficiency for weak beta, such as gas counting, and the carbon may be converted to acetylene, since there is twice as much carbon per unit volume of gas in this form as there is with CO_2. Liquid scintillation counting after conversion of the carbon to an organic form is another good method. Strict precautions to eliminate background radiation are essential. Anti-coincidence circuitry excludes interference from cosmic rays, and shielding, to cut out local gamma, is preferably made of iron rather than lead, as the latter may contain lead-210; pre-nuclear age iron, too, is preferable, and there is some demand for the guns of old warships for this purpose, as they make excellent shields. It is also helpful to use glass free from potassium, i.e. from ^{40}K.

Tritium is present in all atmospheric water vapour, and the principle by which it is used for age determinations is exactly the same as for carbon dioxide. The age range is less, because of the much shorter half-life, and technically it is even more difficult to count than carbon-14: the radiation is much weaker, and the specific activity much less—a gallon of water gives only some 600 disintegrations per second at best—so that samples must be concentrated. The method is useful, e.g. with subterranean water masses. Its accuracy was originally checked with vintage wines, whose water was of known age—it is possible to think of better things to do with vintage wine, but it is difficult, if not impossible, to think of a better source of water whose age is both fairly great and accurately known!

REFERENCES

ABDEL-MALEK, A. A., TANTAWY, A. O., and WAKID, A. M., (1966). 'Eradication of *Anopheles pharoensis* Theobald by the Sterile Male Technique using Cobalt-60', *J. econ. Ent.*, **59**, 672–677.

ALPER, T., (1968). 'Oxygen Effect—Pertinent or Irrelevant to Clinical Radiotherapy', *Br. J. Radiol.*, **41**, 71.

ANDREWS, J. R., (1968). *Radiolobiology of Human Cancer Radiotherapy*, W. B. Saunders, Phila., 271 pp.

ARDRAN, G. M., and CROOKS, H. E., (1968). 'Checking Diagnostic X-ray Beam Quality', *Br. J. Radiol.*, **41**, 193–198.

BARKER, H., (1967). 'Radiocarbon Dating of Collagen in Bones', *Nature, Lond.*, **213**, 415–417.

BERGER, R., and LIBBY, W. F., (1969). 'Equilibration of Atmospheric Carbon Dioxide with Sea Water—Possible Enzymic Control of the Rate', *Science, N.Y.*, **164**, 1395–1397.

BOWMAN, R. C., (1968). *Irradiated Foods—Physical Biochemical, and Organoleptic Considerations*, Project Study Report, Department of Building and Public Health, University of Aston in Birmingham.

BROERSE, J. J., and BARENDSEN, G. W., (1967). 'Measurements of the Biological Effects and Physical Parameters of Neutrons from the $^{9}Be(^{3}He, n)^{11}C$ Reaction—Possibilities for Fast Neutron Radiotherapy', *Int. J. Radiat. Biol.*, **13**, 189–194.

CORNWELL, P. B., (Ed.), (1966). *Entomology of Radiation Disinfestation of Grain*, Pergamon, Oxford, 236 pp.

DEELEY, T. J., and WOOD, C. A. P. (Ed.), (1967). *Modern Trends in Radiotherapy*, Butterworths, London, 382 pp.

ENTZIAN, W., SOLOWAY, A. H., RAJU, R., SWEET, W. H., and BROWNELL, G. L., (1966). 'Effect of Neutron Capture Irradiation on Malignant Brain Tumours in Mice', *Acta radiol.*, **5**, 95–100.

FAO/IAEA, (1966). *Proc. int. Symp. Radiation Preservation Foods, Karlsruhe*, STI/PUB 127, International Atomic Energy Agency, Vienna.

FIELD, S. B., JONES, T., and THOMLINSON, R. H., (1968). 'Relative Effects of Fast Neutrons and X-rays on Tumour and Normal Tissue in the Rat. II. Fractionation', *Br. J. Radiol.*, **41**, 597–607.

HAYMOND, H. R., HAIG, P. V., and KIMBALL, W. R., (1966). '*In Vivo* Thyroid Autoradiography', *J. nucl. Med.*, **7**, 620–624.

HUANG, W. H., and WALKER, R. M., (1967). 'Fossil Alpha-Particle Recoil Tracks For Age Determination', *Science, N.Y.*, **155**, 1103–1106.

IAEA, (1967–1). Radiosterilisation of Medical Products, Pharmaceuticals, and Bioproducts, *IAEA tech. Rep.*, No. 72.

IAEA, (1967–2). *Microbiological Problems in Food Preservation by Irradiation*, International Atomic Energy Agency, Vienna, 148 pp.

IAEA, (1968). *Radiation, Radio-Isotopes and Rearing Methods in the Control of Insect Pests*, International Atomic Energy Agency, Vienna, 148 pp.

JOSEPHSON, E. S., BRYNJOLFSSON, A., and WIERBICKI, E., (1968). 'Engineering and Economics of Food Irradiation', *Trans. N.Y. Acad. Sci.*, **30**, 600–614.

KRAYBILL, H. F., and WHITEHAIR, L. A., (1967). 'Toxicological Safety of Irradiated Foods', *A. Rev. Pharmac.*, **7**, 357–380.

LIBBY, W. F., (1955). *Radiocarbon Dating*, 2nd edn, University of Chicago Press, Ill., 175 pp.

MARCHAL, M., WEILL, J., MARCHAL, M., QUENEE, R., MANSUI, G., and LEPINAT, A., (1966). 'New Method of Medical and Biological Investigation', *C.r. hebd. Séanc. Acad. Sci., Paris*, **263**, 296–299.

MEREDITH, W. J., (1968). *Radium Dosage—the Manchester System*, 2nd edn, E & S Livingstone, London, 170 pp.

REBER, E. F., RAHEJA, K., and DAVIS, D., (1966). 'Wholesomeness of Irradiated Foods—An Annotated Bibliography', *Fedn Proc.*, **25**, 1529.

RINEHART, R. R., and RATTY, F. J., (1967). 'Mutation in *Drosophila* Cultured on Irradiated Food or its Components', *Int. J. Radiat. Biol.*, **12**, 347–354.

ROSEN, L., (1969). 'The Los Alamos Meson Factory', *Sci. Jnl*, **5A**, (July), 39–45.

RUSZNYAK, I., (1968). 'Biological Effects of Ultraviolet Radiation Generated Within Living Organisms', *Experientia*, **24**, 863.

SCHUBERT, J., (1969). 'Irradiated Foods—Chemical, Biological, and Public Health Aspects', *Hlth Phys.*, **17**, 376.

S. I., (1969). *Food (Control of Irradiation) (Admenment) Regulations*, No. 1039, H.M.S.O., London.

VAETH, J. M., (Ed.), (1968). 'Hyperbaric Oxygen and Radiation Therapy of Cancer', *First Annual San Francisco Cancer Symp.*, S. Karger, Basle, Switzerland, 210 pp.

VANDENBRENK, H. A. S., (1968). 'Hyperbaric Oxygen in Radiation Therapy', *Am. J. Roentg.*, **102**, 8–26.

Chapter 6

HEALTH PHYSICS

6.1 THE UNHEALTHINESS OF IONISING RADIATION

6.1.1 The definition of health physics

Health physics is the study of the principles and practice of the protection of man and his environment against the harmful effects of ionising radiations (Rees, 1967). Its aim is prevention rather than cure, seeking to protect from these effects rather than to repair them after they have been suffered. It has several aspects. Radiation *dosimetry* is the quantification of radiation, its measurement, and the relation of radiation dose to biological effect. Through this, guide-lines may be laid down for maximum permissible doses, concentrations, levels, and *body burdens,* i.e. the highest doses which carry a negligible probability that serious injury will result, and the maximum amounts of radionuclides that may be present in water, food, air, on surfaces, and in the body. Radiological safety deals with the correct methods for safe handling of radiation sources and safe disposal of radioactive waste. Personal monitoring of radiation workers keeps a check on the doses they receive, and guards against maximum permissible doses being exceeded. Area monitoring assesses the safety of an area with respect to the spread of activity within it, and is carried out both as a routine check and following an accident. It may be small-scale, as of the bench and apparatus at the end of a radiotracer experiment, or large-scale, covering several square miles, as might happen after an 'incident' at a nuclear reactor plant. Any active areas thus revealed are then, if possible, decontaminated. Finally, the health physicist has something of a public relations job, to educate in these matters and to dispel the mystery that sometimes surrounds radiation.

6.1.2 Nature of the danger

The greatest danger with ionising radiation is that it is so insidious. It cannot be sensed by any of the five natural senses—sight, smell, hearing, taste, or feeling (though it is possible for very high intensities to be detected by the retina of the eye)—and specialised equipment is necessary to reveal its presence, so that exposure can easily take place without awareness of it at the time. The hazard is accentuated by the fact that biological radiation damage is done at the instant of exposure, maybe irreversibly, and it is too late to do much about it by the time any symptoms appear. This is why prevention is so vital in health physics, for cure is difficult or impossible. Clearly, the dangers from ionising radiation are quite different in nature from those encountered more commonly, and it is therefore vitally necessary to exercise great care in using it (Section 6.6).

A legitimate question is of course that, since radiation is so dangerous, would it not be better to shun it altogether? This may be desirable, but it is not possible: first, like many other things that make good servants but bad masters, radiation is now too useful and beneficial to be discarded simply because of a certain element of danger in its use, and indeed, if handled with respect, it is no more dangerous than anything else. In addition, the full potential of radiation as a tool in Man's service has not yet been realised, and further study is necessary to do this and to do it safely. Secondly, radiation exists as an established fact of present-day life, whether we like it or not; its action on living things is by no means fully understood, and work must go on until such an understanding is gained. Besides being valuable for its own sake, this knowledge is essential for the more accurate determination of maximum acceptable radiation doses, body burdens, and concentrations, and for the treatment of over-exposure to radiation. It is also essential for the safe development of further uses of radiation and for safe disposal of the radioactive waste arising from the use of isotopes. Release of radiation into the biosphere creates a debt which has to be paid off by future generations, and the results of atomic bomb explosions and weapons tests since 1945 are an aweful warning against releasing radiation without control and without thought for the biological consequences.

All radio-isotopes are dangerous, but some are more dangerous than others, and they are divided into four classes of radiotoxicity, as follows:

Class 1 is very highly toxic, and comprises the alpha emitters and strontium-90.

Class 2 is highly toxic: of biological interest are sodium-22 and chlorine-36, and isotopes that concentrate selectively in particular organs, such as iron-59, iodine-131, calcium-45 and barium-140.

Class 3 is moderately toxic, and includes most tracers, e.g. phosphorus-32, sulphur-35, potassium-42, carbon-14, etc.

Class 4 is of low toxicity, and includes tritium.

Maximum permissible figures for concentration of radionuclides in water and air (MPC), for surface contamination (MPSC), and for body burdens (MPBB), are calculated primarily on a basis of toxicity, according to the above classification, and taking account of the factors affecting radiation dose from an internal source (Section 6.1.4). Specimen figures are:

	Carbon-14 (Class 3)	*Plutonium-239 (Class 1)*
MPC in water	2×10^{-2} μCi/ml	1×10^{-4} μCi/ml
MPC in air	4×10^{-6} μCi/ml	2×10^{-12} μCi/ml
MPBB	3×10^{2} μCi	1×10^{-2} μCi (0·0005 μg)

The MPBB for polonium-210 is 0·03 μCi, equivalent to a mass of $6\cdot8\times10^{-12}$ g, so that, weight for weight, it is some 10^{11} times more toxic than cyanide. The MPSC is 10^{-3} μCi/cm² for benches, etc., and 10^{-4} μCi/cm² for the body, for isotopes other than alpha emitters, for which the maximum limits are a tenth of these figures.

The chemical form of an isotope can greatly increase its toxicity, the best example being tritium: it is ten thousand times more toxic as tritiated water than as tritium gas, and it is worth noting that HTO is readily absorbed through the skin. Tritiated water may exchange in the body sufficiently to label organic molecules significantly, giving whole-body doses about half as much again as those determined on the assumption that the tritium stays in body water until it is excreted (Evans, 1969). The hazard from tritium is increased by the extreme softness of its radiation, which makes the detection of contamination with it very difficult. Estimation of the body burden of tritium in those working with it is easier, more informative, and more satisfactory than a monitoring survey of the laboratory: determination of tritium in exhaled water is a valid method for this purpose, and it may be more acceptable than the customary method of urine sampling (Chiswell and Dancer, 1969). Tritiated thymidine and similar molecules are even more hazardous than HTO, likewise DNA labelled with carbon-14, since they can readily be incorporated into the body's reproductive machinery, and they should be treated with the same respect as Class 2 nuclides. Vennart (1969) discusses the radiotoxicity of tritium and carbon-14 compounds.

The amounts of isotope to be handled in the laboratory are governed by the radiotoxicity of the isotope and by the quality of the laboratory.

Grade A is the 'hot' laboratory, purpose-built for high-activity work, including remote-handling facilities, changing rooms, showers, etc.

Grade B is specifically designed for isotope work, but to a less-exacting standard than Grade A.

Grade C is a high-quality chemistry laboratory.

The maximum recommended amounts of activity in each grade are:

Grade	A	B	C
Toxicity			
Class 1	10 mCi	1 mCi	10 μCi
Class 2	100 mCi	10 mCi	100 μCi
Class 3	1 Ci	100 mCi	1 mCi
Class 4	10 Ci	1 Ci	10 mCi

These figures apply to normal chemical operations, and modifying factors may be applied according to the type of work, as follows:

Storage	×100	Complex wet work, with risk of spills	×0·1
Simple wet operations	×10	Dry and dusty work	×0·01

Health hazards from radiation arise through *irradiation*, either external or internal, from sources outside or inside the body, and through *contamination*. The two must be carefully distinguished: a highly active sealed source is a radiation hazard, but not a contamination hazard, whereas an open source in the tracer laboratory presents little or no radiation danger, but a high risk of contamination.

Specialised problems are also involved in the handling of fissile nuclides, though they will not be considered in detail here; they present a unique combination of hazards—high radiotoxicity, intense radiation, and criticality—making their control a challenge to the health physicist (Denham, 1969). There is the possibility of a *criticality incident* with sub-critical quantities which together exceed the critical mass; if these should come together, an uncontrolled chain reaction, which happens extremely quickly, may result in an explosion and the release of a vast amount of radiation (Section 2.4.4) and contamination.

6.1.3 External irradiation

The greatest hazard, whole-body irradiation, arises from the penetrating radiations, x-rays, gamma rays, and neutrons. Large sources, such as those used in radiotherapy, nuclear physics, and isotope production and processing, are likely to affect all parts of the body more or less equally, whereas smaller sources, such as are used in radiotracer experiments, irradiate locally. Although smaller sources will deliver some radiation to all parts of the body, the inverse-square law operates, and the highest dose will be received in the immediate neighbourhood of the source, falling off rapidly as the distance from the source increases.

Beta radiation presents little external hazard, inasmuch as none but the hardest can penetrate more than about a centimetre deep, and it affects only the superficial layers of the body. Irradiation is local, rather than whole-body, and most commonly affects the hands and fingers, e.g. when a small source is being manipulated in a radiotracer experiment. Some method of fingertip or hand dosimetry may therefore be very useful, for a hard-beta source can give a heavy dose to the extremities with little or no effect on the rest of the body. The eye lens is vulnerable to beta radiation, so it is inadvisable to look too closely at a strong source.

There is virtually no hazard from alpha radiation outside the body, for the fact that alphas are absorbed within a very short distance means that they are stopped by the outer layers of the epidermis or by an inch or two of air (Section 2.4.3).

Contamination of the body surface is a source of external irradiation.

6.1.4 Internal irradiation

Internal irradiation arises from an external source of penetrating radiation, or from introduction of a radiation source into the body, either deliberately (medical use) or accidentally (which may be regarded as internal contamination). The effect on the body of an internal source is qualitatively similar to that from external irradiation, but the danger from an internal source of radiation is very much more serious than from an external one of the same strength, in that it is impossible to move away from the source or to shield the body from it, and the source is much nearer the sensitive target areas of the body.

Radio-isotopes may enter the body by various routes: by mouth, in food, drink or by inhalation; by injection; or by absorption through a wound or the skin (percutaneous). The fate of the isotopes then depends on their nature and properties, and they are not necessarily incorporated into the tissues—insoluble or indigestible material entering the alimentary canal will pass straight through without absorption. Retention of particles in the lungs depends on the anatomy and physiology of the lung—dogs' and cats' lungs are not typical of mammalian lungs—and on particle size. The largest particles are filtered in the nostrils and those down to about 10 μm are trapped in the upper part of the respiratory tract. Particles down to 5 μm are caught lower down in the film of mucus lining the bronchi and nasal passages, in which they may eventually be expelled from the body, or transferred to the alimentary canal. Particles smaller than about 5 μm are completely retained in the lungs, down to about 0·4 μm, below which they become more akin to a vapour and may be exhaled. Some particles may be taken into the body from the lungs by phagocytosis. It may be noted that in aerosols, a common form of airborne contamination, particles of about 3 μm form the largest proportion by weight. Inhalation is a more important route than might be realised, since the volume of air breathed in per day is far greater than the volume of food consumed.

The radiation dose given to the tissues by an internal source depends on the type and energy of the radiation, and on its persistence and distribution within the body.

Type and energy of the radiation

The extent of the damage caused to the tissues depends on the rate at which the radiation loses energy to the tissues per unit of path length, the *linear energy transfer* (LET), which determines the intensity of ionisation produced. For particulate radiation, the LET varies with the square of the charge and inversely with the velocity. Thus alpha particles, with a double elementary charge and low velocity, have a very high specific ionisation, of many thousands of ion pairs per centimetre, and cause very intense local damage immediately round the source. In complete contrast to their negligible external hazard, alphas are extremely dangerous internally, even though the distance to which they penetrate is very small, for their considerable energy is dissipated within an exceedingly small volume of tissue, causing great local damage. In addition, most alpha emitters are members of a decay chain and, besides the alpha radiation, beta and gamma arising from other members of the chain may also be present. Beta particles present somewhat less of an internal hazard than alphas.

Consideration of how the energy of particulate radiation from an internal source affects the tissue dose yields the apparent contradiction that the intensity of irradiation increases with decreasing energy, though the explanation is quite simple, and follows from the geometry of the situation. Assume that there are two sources, emitting particles of energy e and $2e$ which penetrate for distances r and $2r$ respectively. The energy emitted will be absorbed in a spherical volume around the source, whose radius will be r or $2r$. If the sources emit equal numbers of particles, the intensities of irradiation will be in the ratio:

$$\frac{e}{\frac{4}{3}r^3} \quad \text{to} \quad \frac{2e}{\frac{4}{3}(2r)^3} \quad \text{or} \quad 4 \text{ to } 1$$

If the sources emit the same total amount of energy, the intensities of irradiation become as 8 to 1. Thus a low-energy beta-emitter presents a greater internal hazard than a higher energy one, since it produces a much higher intensity of irradiation.

Gamma radiation from an internal source will irradiate the whole body; as with localised external irradiation, the inverse-square law means that the highest dose is received in the immediate locality of the source. Some radiation will escape from the body altogether, and can be picked up by an external detector; it is therefore possible to undertake *whole-body monitoring* for the presence of internal gamma emitters, whereas it is not possible with internal beta emitters.

The energy absorption per röntgen (Section 6.2) of gamma or x-radiation is roughly constant for all tissues for radiation energies above about 0·2 MeV, at about 100 erg per g tissue per röntgen (1 erg$=10^{-7}$ J). Below this, the energy absorption in hard tissue (bone) increases sharply, reaching 500 erg per g per röntgen for 0·01 MeV radiation, but there are only a few isotopes of practical importance in this region.

Persistence of the radiation

The persistence in the body is governed by the half-life of the isotope and by the rate at which it is excreted. Thus carbon-14, with its long half-life and soft beta radiation, would appear to be quite dangerous if it should be ingested, were it not for the fact that it normally has a rapid turnover in the body and is excreted fairly quickly as carbon dioxide. On the other hand, if it should become fixed within the body in a form which is not in equilibrium with the metabolic pool, then it will be that much more hazardous. For example, glycine labelled with carbon-14 is not accumulated to any appreciable extent in the adult, but it is incorporated into the structural elements (bone and cartilage) of growing subjects, where it will remain for a long time. Tritiated thymidine is also highly dangerous, in spite of the rapid turnover of most body hydrogen, since it fixes in DNA.

The biological half-life of a labelled compound, $b_{1/2}$, and the radioactive half-life of the isotope, $t_{1/2}$, may be combined into an expression for the effective half-life of the radiation within the body, $E_{1/2}$, given by:

$$\frac{1}{E_{1/2}} = \frac{1}{t_{1/2}} + \frac{1}{b_{1/2}}$$

Thus, provided that they are freely exchangeable within the body and are not fixed in a non-exchangeable form, the effective half-lives of carbon-14 and tritium are 35 d and 19 d respectively, compared with their radioactive $t_{1/2}$'s of 5570 and 12·3 years. The difference is less apparent with sodium-24 and phosphorus-32, whose $E_{1/2}$'s are only slightly shorter than their $t_{1/2}$'s. This offers a method for determining the biological half-life of a compound: it can be labelled and fed to an animal, and the rate at which it disappears from the body is measured; knowing $t_{1/2}$, $b_{1/2}$ can then be calculated.

Distribution of the isotope within the body

Concentration of a radioactive compound or element presents a greater hazard than if it is evenly spread throughout the body, e.g. with iodine, iron, calcium, and strontium, which accumulate in certain tissues. Wide distribution of an isotope is not necessarily synonymous with even distribution: phosphorus is spread over the whole body, but it is present in greater concentration in lymph nodes and kidneys than elsewhere. Thus, if a compound or element is being used and little is known about its distribution in the body, it should be handled cautiously.

This leads to the concept of the ***critical organ*** for an isotope or a labelled compound, which is the organ suffering most damage through ingestion of that isotope or compound. Thus the critical organ for iodine-131 is the thyroid gland, which concentrates iodine far more than any other organ. The critical organ for a short-lived isotope may not be the one in which it ultimately concentrates, for it may well have decayed away so much by the time it gets there that it will cause less damage to this organ than it has done to others (e.g. stomach or intestine) en route. Where an isotope is distributed widely and uniformly throughout the body, the whole body is taken to be the critical organ. The maximum permissible body burdens of particular isotopes are worked out on the basis of the damage they cause to their critical organs.

6.1.5 Contamination

Contamination can be regarded as radioactivity in the wrong place, in the same way that dirt may be defined as matter in the wrong place, and like dirt it too should be cleaned up promptly and not left lying about. The chief risk with contamination is that its existence may not be suspected, so that it may give a large radiation dose before being discovered, or it may be taken into the body unawares. Apart from presenting a health hazard, contamination may be a nuisance in experimental work, especially where low activities are concerned, since it increases background. The danger of contamination arises to the greatest extent with unsealed (open) sources, small amounts of which can easily be left behind when the main bulk is cleared away, or be transferred to other objects. Sealed sources normally present only a radiation hazard, but if the sealing is broken, so that active material can escape, then there is a risk of contamination. Several standardised methods have been laid down for testing sealed sources for leakage, e.g. by wiping and counting the swab, or by immersion in water and counting the activity extracted. Transferable (removable) and non-transferable (fixed) contamination may be distinguished: the latter is firmly attached to the contaminated object, and is an external radiation hazard only, whereas the former is easily removable and is a potential source from which isotope may get into the body. Fixed contamination on walls and fittings must be shielded and labelled clearly.

6.2 DOSIMETRY: UNITS OF RADIATION DOSE

The biological effect of radiation should ideally be measured in biological terms, but it is difficult or impossible to do this since the biological effect of radiation depends on many factors. Radiation units are therefore defined in physical terms, and biological effects are (if possible) correlated with these. Many units are found in dosimetry, because of the above-mentioned difficulty, but the recommendations of the International Commission on Radiological

Units (ICRU, 1968) have helped greatly to resolve it, and they are now widely accepted.

The original unit, which is still in use, is the *roentgen* or *röntgen*, abbreviated to R (small r should not be used). It expresses the quantity of ionisation produced by radiation in air, or the capacity of radiation to ionise air. It is defined as: The exposure dose of x- or gamma-radiation such that the associated corpuscular emission per 0·001293 g (1 cm^3) of dry air produces ions carrying 1 electrostatic unit (e.s.u.) of electric charge of either sign. This is equivalent to $2{\cdot}58 \times 10^{-4}$ C per kg of dry air, or to the production of $2{\cdot}1 \times 10^9$ ion pairs per cm^3 of air. It becomes more meaningful when it is expressed in more practical terms, as: The quantity of x- or gamma-radiation producing an energy dissipation of 87·6 erg per g of air. This amount of energy is equivalent to $5{\cdot}18 \times 10^7$ MeV, or $1{\cdot}98 \times 10^{-6}$ cal. (1 erg $= 10^{-7}$ J and 1 cal $=$ 4·2 J.) The röntgen is by definition limited to electromagnetic radiation (strictly speaking, to that of energy below 3 MeV) and to air—different amounts of energy will be dissipated if other materials are exposed to 1 röntgen of radiation, e.g. 98 erg/R of medium-energy radiation are dissipated per g of soft tissue. The röntgen is a unit of radiation *exposure*, not of dose; exposure rates are expressed as R/h, mR/min, etc.

A more convenient unit would express energy absorption in tissue rather than in air, and do so without reference to the energy and type of radiation. The rep (röntgen equivalent physical) was devised for this purpose, but it has now been replaced by the rad (radiation absorbed dose), introduced by the ICRU in 1953. The rad is a unit of *absorbed dose* and is equal to 100 erg absorbed per g of a specified material from any ionising radiation. From the figure above it will be seen that, for gamma radiation, the absorbed dose per röntgen is 0·98 rad in soft tissue, so that here the röntgen is numerically almost exactly equivalent to the rad, although the two are defined on different bases. One rad is equivalent to $6{\cdot}25 \times 10^7$ MeV, or $2{\cdot}39 \times 10^{-6}$ cal.

Different biological effects may be produced by the same dose in rads of different types of radiation, because an important factor in the biological effect of radiation is not merely the total energy absorbed, but the *linear energy transfer*, i.e. the rate at which energy is transferred to the tissues per unit of path length, or the intensity of ionisation produced. This leads to the concept of the *relative biological effectiveness* or *quality factor* (RBE or QF) of a radiation, which expresses how effective a radiation is in producing a particular change. It is defined as the ratio of the dose of gamma radiation needed to produce a given effect to the dose of the specified type of radiation needed to produce the same effect (one source quotes gamma radiation from cobalt-60 as the reference; another quotes 200 kV x-rays). The RBE of all gamma radiation and most beta particles is generally taken to be 1, increasing to about 2 for soft betas, so that in practice röntgens, rads, and rems (*see* below) for many common radiations are all numerically equal, within the $\pm 10\%$ accuracy which is all that is normally required in health physics measurements. The RBE value is about 3 for thermal neutrons, and 10 for protons, alphas, and fast neutrons. Heavy ions have RBE up to 20. The term RBE is now limited to experimental radiobiology, having been replaced in health physics by QF. The unit of *dose equivalent*, the rem (röntgen equivalent man or röntgen equivalent mammal) takes RBE (QF) into account: it is equal to the dose in rads multiplied by RBE (or QF). It can be regarded as the unit of biologically effective dose. When the QF is 1, the rad (in air) is equal to the dose equivalent in rem which would be received in soft tissue at the point of measurement. A further term, *dose distribution factor*, is used to express the modification of the biological effect arising from the non-uniform distribution of an isotope deposited within the body.

A comparison of the energy relationships involved in radioactivity with those of other atomic changes is illuminating, revealing as it does the tremendous power of ionising radiation. Chemical reactions involve energy changes of about 10 eV per atom, whereas those in radioactive transmutations are two to five orders of magnitude greater, in the keV to MeV range. Thus a gram-atom of carbon burning to CO_2 releases 94 000 cal, but transmutation of a gram-atom of carbon-14 to nitrogen releases 3×10^9 cal. Yet, on the other hand, the energy content of 500 rad of x-rays, a fatal whole-body dose, is equivalent to only 5×10^4 erg per g of tissue, compared with the $4{\cdot}2 \times 10^7$ erg of heat energy needed to raise the temperature of 1 g of water through 1°C.

6.3 DOSIMETRY: CALCULATION OF RADIATION DOSES

Radiation dosimetry is basic both to an assessment of the risks from man-made radiation in the environment as compared to naturally occurring radiation, and to the safe handling and use of radiation sources. It links the activity of a source to its biological hazard. Dose calculations, complemented by actual measurements, are important in planning experiments which involve approaching or handling a large source, by giving an estimate of the maximum time for which this may be done without danger. They are also important in organising decontamination operations after an accident with a large source, to determine how long a worker may be in the contaminated area without receiving more than his permitted dose (Section 6.5.4), and how much radiation was received by anyone in the vicinity at the time the accident occurred. Dose calculations must take account of the decay scheme of the isotope (Section 2.3.1).

Electromagnetic radiation

Dose calculations are mostly concerned with gamma sources, since penetrating radiation presents most problems. Since 1 R is then numerically almost exactly equivalent to 1 rad for soft tissue (Section 6.2), figures for exposures in R and dosages to the body in rad are for practical purposes the same.

The exposure rate (R per unit time) from electromagnetic radiation (and hence the dose) is roughly proportional to the energy from about 0·06 MeV upwards; below this point, the exposure increases sharply as the energy decreases. However, few isotopes have radiation below 60 keV, and none are of direct biological interest. The exposure–energy relationship is slightly curved, hard radiation giving proportionately less exposure than that of lower energy. Accurate estimates can be made by reference to a graph, or to tables of the *specific gamma constant* or *emission* (formerly known as the *k factor*), the exposure rate in R/h at 1 cm from a 1 mCi source. These tables are very useful, especially with isotopes having complicated decay schemes where calculations are tedious.

Radiation exposures from gamma radiation of energy between about 0·3 and 3 MeV may be calculated by means of an equation connecting the flux, N, (the number of photons passing through each cm^2 per s) and the linear energy transfer, L, (in MeV/cm) of the radiation at the point in question; the exposure rate is $5{\cdot}76 \times 10^{-5} LN$ R/h. Calculations of L and N can become a little complicated, and a simpler equation is almost invariably used in practice; it has the advantage of slightly over-estimating the exposure rate from radiation above 1 MeV energy. If C is the curie strength of the isotope and E is its *total* gamma energy in MeV, then the exposure rate D in R/h at 1 ft (0·3 m) $= 6\,CE$. At 1 m, $D = \frac{1}{2}CE$. Thus, cobalt-60 emits two gamma rays per disintegration, and for a 1 mCi source $D = 6 \times 10^{-3} \times (1{\cdot}33 + 1{\cdot}17) = 0{\cdot}015$ R/h at 1 ft. Potassium-42 emits a gamma ray of 1·52 MeV in 18% of its disintegrations, and the E term in the equation would here become $(1{\cdot}52 \times 0{\cdot}18)$. The inverse-square law is applied to calculations at other distances.

Beta radiation

Beta radiation is common in tracer work, but it is not often necessary to calculate external doses from it, as it is so easy to shield it completely, unlike gamma. Calculations may be required, for example, in the dispensing of a small aliquot from a large stock, which must of necessity then be approached closely with at least some of its shielding removed. If C is the curie strength of the source, then the dose rate $D = 300\ C$ rad/h at 1 ft. As with gamma, the inverse-square law again applies for calculation of dose rates at other distances. The formula assumes that the radiation is not attenuated by absorption in the air it passes through, and therefore errs on the safe side by overestimating the dose. A factor T to account for transmission losses can be added to obtain a more accurate estimate of the dose, though it is rather tedious to calculate. T is equal to $2^{-x/d_{1/2}}$, x being the amount of air in mg/cm^2 from the source to the exposure point: at 1 ft x is 40 mg/cm^2. The term $d_{1/2}$ is the half-thickness for the beta radiation in mg/cm^2; it is roughly equal to $46(E_{max})^{3/2}$.

Internal irradiation

The dose rate from isotopes within the body can be calculated without difficulty from the number of particles emitted, their energy, and the fact that 1 rad $= 6{\cdot}25 \times 10^7$ MeV; in rad/d it is $50{\cdot}7\ AE/W$ where A is the quantity of isotope in microcuries and W the weight of tissue in which it is distributed. E is the average energy, E_{av}, for beta particles, and the total energy for alpha and gamma. The dose from gamma may be over-estimated since it may not be completely absorbed within the body.

6.4 DOSIMETRY: MEASUREMENT OF RADIATION DOSES

Radiation doses may be measured by a number of different instruments, which vary both in the principle on which they operate and the purpose for which they are designed (Aglintsev, 1965). Radiation monitors are used for survey work; and the film badge, ionisation chamber, and thermoluminescent and photoluminescent dosimeters are used for low dose measurements, particularly in personal dosimetry. Calorimetric, colorimetric, and chemical methods are applied in measuring high doses.

6.4.1 Radiation monitors

Radiation monitors give a reading in approximate proportion to the intensity of the radiation they receive, in terms either of count rate (counts/min or counts/s), or dose rate (mrad/h), so that they are also known as *ratemeters.* The former term is more applicable when surveying an area for freedom (or otherwise) from contamination, as in the clearing-up stage after a tracer experiment, or in the counting of low-activity sources giving only a few counts above background. An audible indication may also be provided, each ionising event that is detected giving a click in a loudspeaker, which is very useful when tracking down 'hot spots' of contamination on a surface. A monitor indicating dose rate is used to assess the safety of an area for working, i.e. to determine how long a person may stay there without receiving more than the permitted amount of radiation.

Monitors and dosemeters and ratemeters should be tested once a year, under the Ionising Radiations (radiation dosemeter and dose rate meter) Order 1961, SI No. 1710: their indicated scale reading is checked against the true dose rate from a source of known activity, and the two are graphed. Through this calibration, an instrument reading in terms of count rate may be used for dose monitoring.

Radiation monitors are based on an ionisation chamber or G–M detector for gamma and beta radiation, or a scintillation detector for alpha, plus a counting circuit very like a scaler–timer set except that an *integrating circuit* is added at the end. This records the radiation received over a period of time, then works out the average per unit time, and presents it as the reading on the dial. The length of time for which the radiation is recorded is the *time constant;* it is usually variable at will. The instrument operates continuously—successive recording periods overlap and are not placed end-to-end—so that the reading does not stay in place for the duration of the time constant and then move to a new value, but fluctuates constantly; the true reading of the instrument is taken as the mid-point of the range over which it is wandering, as judged by eye. (Some ratemeters feed in to a strip-chart recorder, from which the reading can be estimated more accurately at leisure.) The instrument reading random fluctuations arise from the nature of radiation, and are greatest with low fluxes. They are more and more averaged out, and the reading of the instrument thereby steadied, by increasing the time constant.

Even though a G–M tube gives a constant response whatever

the energy of the particle or photon that 'fires' it, its use for dosimetry of gamma is valid because, as the dose received from gamma radiation increases in proportion to energy (Section 6.3), so does the efficiency with which the tube detects it (above about 0·3 MeV), so that its response then becomes roughly proportional to the dose rate. It may be noted that radiation monitors may not be sensitive to small dose rates and that, at very high fluxes, G–M tubes may jam completely, registering nothing at all.

More-specialised monitors are used for checking airborne contamination; a pump is used to suck a stream of air through them, so that more radiation may be detected.

Neutron monitors are also specialised in design. They embody elements of high neutron-capture cross-section (Section 2.4.4) (boron or cadmium) and a proportional counter to detect the ionising particles released when the neutrons interact with these elements. Such monitors are surrounded by a moderator such as polythene that slows down the neutrons and increases the chance of interaction.

6.4.2 Photographic film dosimetry

The film badge is used widely, if not universally, for *personal dosimetry of beta and gamma radiation.* It consists of a standard dental x-ray film, $1\frac{1}{2}$ in $\times$ 1in, worn in a holder incorporating eight filters (Figure 6.1). An open window, and two thicknesses of plastic, interpret the beta exposure; aluminium–copper alloy and (tin+lead) filters absorb respectively softer and harder electromagnetic radiation; and a (cadmium+lead) filter evaluates exposure to slow and thermal neutrons. A piece of lead at 90° to the lead filters, on the outside, eliminates edge effects on the film from soft gamma or x-irradiation. All these filters are duplicated at the front and back of the holder, but the eighth filter, a 0·4 g strip of indium, is at the front only. It enables x- or gamma-exposure from front and back to be differentiated, though its main function is to detect exposure in a criticality incident; the neutrons released induce activity in the indium, and the medium-hard beta emitted can be picked up at once by a radiation monitor. (Heard and Jones, 1963.) The presence of the several filters means that the different types and energies of radiation are attenuated to varying degrees. The film is not therefore uniformly blackened by the radiation, but shows a pattern according to the extent of this attenuation, and the amount of blackening in the different parts enables an assessment to be made by reference to standards of the dose received from different radiations (Figure 6.2).

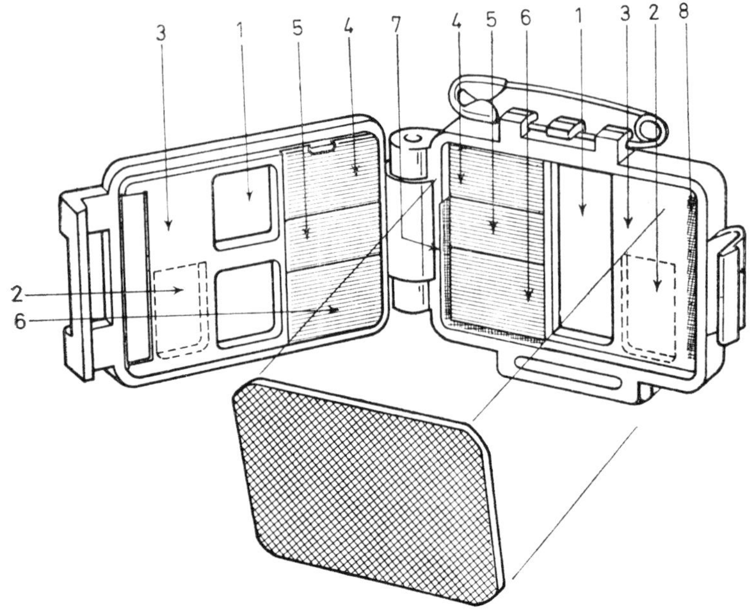

Figure 6.1. Film badge holder. (From Heard and Jones, 1963. Courtesy U.K.A.E.A.)

The filter types are as follows:

1. *Window*
2. *50* mg/cm² *plastics*
3. *300* mg/cm² *plastics*
4. *0·040* in *Dural*
5. *0·028* in Cd+*0·012* in Pb
6. *0·028* in Sn+*0·012* in Pb
7. *0·012* in Pb *edge shielding*
8. *0·4* g In

Though more-recently developed dosimetric methods may be better than film in some ways, film too has its own particular advantages, and it is suggested that its development should be kept up (Becker, 1966; Geiger, 1968; Landauer, 1968). Thus, although film badges cannot give an immediate reading of dose, they are permanent, and a suspected abnormal reading can be checked, several times if necessary, without erasing it in so doing; this may be of legal value if a worker should claim to have suffered ill-health from over-exposure to radiation.

Different sorts of film badge are used for dosimetry of ***neutrons***. For fast neutrons above 0·5 MeV, a nuclear emulsion (Section 3.4.4) is placed on either side of a polythene 'radiator', with which the neutrons interact: the recoil protons from the interaction are registered in the emulsion, and their number gives a measure of the neutron dose equivalent. Some skill is needed to interpret the results, and sensitivity to low fluxes of neutrons depends on the film being free from fogging from other radiation. Measurement

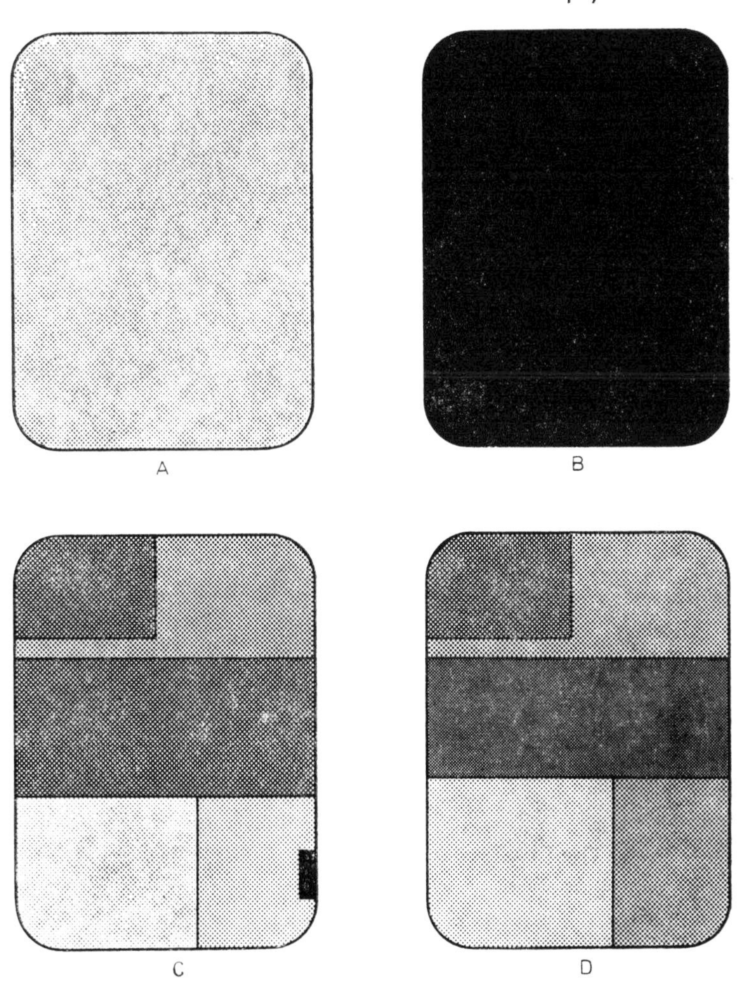

Figure 6.2. Exposed radiation films: A—*less than 20* mrad, *equivalent to unexposed;* B—*50* mrad *of 70* kV *x-rays, with no holder;* C—*20* mrad *of 40* kV *x-rays;* D—*Radium-226, alpha with some gamma. Note effect of metal filters in* C *and* D. *(The number of the film is normally visible across the centre)*

of thermal neutrons below 10 keV utilises a (n, γ) interaction with an element of high capture cross-section: a cadmium filter is placed in front of a standard film dosimeter, which then records the gamma resulting from the interaction. An (n, α) reaction with an element like boron ($^{10}B(n, \alpha)^{7}Li$) may also form the basis for dosimetry. Personal dosimetry of slow neutrons in the 10–500 keV range is not covered by these two methods. A way out of the difficulty is to determine the spectrum of the neutron flux, and extrapolate the results of fast and thermal dosimetry on the basis of the relative proportions of the several types of neutron. Nagarajan and Krishnan (1969) suggest a composite device for monitoring both thermal and intermediate-energy neutrons: a plastic medium, for recording the damage tracks due to neutron interactions, is covered on both sides by a layer of boron (120 μm thick) with cadmium (0·5 mm thick) on the outside. The cadmium at the front of the device extends over only half the area, so that under this area thermal neutrons are cut off and only intermediate-energy neutrons are recorded; the remainder of the device records both types. Figures of 2350 tracks per cm^2 per mrem are reported for thermal neutrons, and 50 for intermediates. Peirson (1968) discusses the approaches and problem in neutron personal dosimetry.

The Radiological Protection Service, organised by the Ministry of Health and the Medical Research Council, operates a film-badge service for both beta–gamma–x-ray and for neutron monitoring; the price of the former is currently £0·15 per film. The films are usually issued for a month, and are then returned to the Service for processing and interpretation, accompanied by a report form (in duplicate) giving the name of the wearer, the serial number of the film, the position of wear, the isotopes or radiations used, and the type of work. The serial number is stamped on the outer wrapper, (where it shows through the open window) and, since the mechanical pressure of the impression penetrates right through to the film itself, a 'deliberate artefact' is created that serves to identify the film after removal from the wrapper. The duplicate copy of the report form is returned in due course with a note of the total dose received by each film, reported to the nearest 10 mrad up to 300 mrad; the accuracy is ± 20 mrad up to 100 mrad and $\pm 20\%$ thereafter. Doses less than 20 mrad cannot be accurately assessed and are taken as meaning no dose significantly greater than the background. The mrad (in air) is equal to the dose equivalent in mrem which would be received in soft tissue at the point of measurement. If the dose is found to be excessive, attention is drawn to it on the report, and suggestions made regarding possible causes and appropriate action, for it is surprising (to the layman) how much information the Service can gather from a square inch of x-ray film. Should there be no satisfactory improvement, a thorough radiation survey of the laboratory is recommended. The address of the Service is: Clifton Avenue, Belmont, Sutton, Surrey.

6.4.3 Ionisation chamber dosimeters

The *quartz-fibre dosimeter* (or so-called pocket dosimeter, because it fits in the pocket like a fountain pen) is a miniaturised version of the electroscope ion chamber detector (Section 3.2.2). It is charged up before use, to zero it, and the movement of the fibre as radiation dissipates the charge is observed through the built-in microscope. A condenser-type pocket ionisation chamber works on the same principle, but does not have a fibre and must be connected to a separate instrument to read the radiation exposure. The charge gradually leaks away whether or not any radiation is recorded and, though the natural leakage is less than 10% in 24 h, the instrument should be recharged at the start of each working day and read at the end of it. More work is therefore involved than with the film badge, but there is the compensating advantage that a reading of dose can be had immediately, whereas films must be sent away for processing. Pocket dosimeters can measure doses up to 50 rad; they are constructed so as to cover one particular range, however, since one instrument cannot satisfactorily record both high and low doses.

6.4.4 Thermoluminescent dosimetry (TLD)

TLD is based on the principle that, on irradiation, some substances can form metastable states that store radiation energy, and release it as light when subsequently heated. Irradiation induces ionisation in the material, and some of the free electrons formed thereby are trapped at imperfections or impurities in it. Heating enables the electrons to escape from these traps, and as they return to normal positions the energy they gained on ionisation is released as light.

The plot of luminescent brightness against temperature as the material is heated at a constant heating rate is the *glow curve*. It contains several peaks, whose number is determined by the number of traps: the depth of the trap, i.e. the energy holding the electron in, determines how much heat energy is needed to free it and therefore the temperature at which that peak appears. Either the whole curve, or a particular peak, is recorded by a photomultiplier; the light output is a measure of how many electrons were trapped, and thus gives a measure of the original ionisation. The amount of light has been found to be a linear function of the dose from 1 mrad to over 10^4 rad. Heating restores the material to its original state, so that it may be used again when cool. It can be recycled indefinitely.

To be suitable for TLD, a material must have electron traps deep enough not to be emptied at ambient temperatures, i.e. its response must not fade with time. Calcium fluoride and calcium sulphate, activated with manganese, have been used, but lithium fluoride activated with magnesium is perhaps the commonest material; being of low atomic number it has the advantage that it is a 'tissue equivalent' substance (Shambon and Condon, 1968). Beryllium oxide (Tochilin, Goldstein, and Miller, 1969) and some special glasses have also been suggested. TLDs can be made very small, and are therefore useful for implantation to determine radiation doses within the body. Their other advantages are great flexibility in design, wide range, and quick re-usability; they can also be made sensitive to thermal neutrons, though they can be permanently damaged by fast neutrons (Kastner, 1969). TLD is more accurate with x- and gamma-radiation than film (Schulman *et al.*, 1960; Hall and Wright, 1968; Spurny, 1969). The Radiological Protection Service offers thermoluminescent dosimeters, at present at the same cost as film.

Related to thermoluminescence is thermally stimulated exo-electron emission (TSEE), the emission of low-energy electrons from the surface of irradiated material (lithium fluoride). TSEE is a function of the surface area, not of volume. It can be measured quite easily, for example by heating inside a gas-flow G–M counter, and has been suggested as a method of dosimetry (Becker, 1969), though TSEE is a very variable quantity and needs much development before it can be considered a useful method.

6.4.5 Photoluminescent dosimetry

Photoluminescent dosimetry makes use of a glass in which ionising radiation causes the formation of stable fluorescing centres which then emit light continuously when excited by ultra-violet irradiation. An example is a silver-activated phosphate glass, but this has a high average atomic number compared with tissue (it contains 10% by weight of barium), which limits its use. Other glasses of lower average atomic number have been developed more recently, and they are now available in different shapes and sizes. Some glasses can measure down to about 20 mrad, and the intensity of fluorescence is a linear function of the dose up to at least 10^4 rad. Glasses can be made which are sensitive to thermal neutrons. Photoluminescent dosimeters have some potential in personal dosimetry, although the minimum dose they can record is higher than with TLD, and they can be re-used only after a long time, if at all; however, this does mean that a more or less permanent record of dose may be accumulated (Becker, 1968).

6.4.6 Calorimetric dosimetry

Calorimetric dosimetry uses the heating effect of ionising radiation as a basis for dosimetry. One rad is equal to $2{\cdot}39\times10^{-6}$ cal (1 rad = 100 erg; 10^7 erg = 1 J; mechanical equivalent of heat = 4·19 J/cal). Thus a dose of 100 rad delivers $2{\cdot}4\times10^{-4}$ cal per g of water, giving a temperature rise of this many °C, which one textbook states to be readily measurable with accuracy under suitable conditions. It seems obvious that these conditions are rather specialised and that calorimetric dosimetry is applicable only to high levels of radiation.

6.4.7 Chemical dosimetry

Chemical methods of dosimetry use the ability of ionising radiation to bring about chemical change. The ideal chemical dosimeter should embody a number of fundamental characteristics: the response to absorption of a given amount of radiation should be independent of changes in chemical concentrations (of starting material, product, or other components of the system), and of the quality and dose rate of the radiation. The analytical determination should be fairly easy, ordinary reagents of AR quality should be usable without any need for special purification, and solutions should be adequately stable.

The perfect system has not yet been found, but the *Fricke–Miller* dosimeter is close enough to the ideal, for all practical purposes, to have been accepted as a secondary-standard dosimeter. Battaerd and Tregear (1966) discuss its practical use and modern developments that increase its range and sensitivity, and Oller, Menker, and Dauer (1969) compare it with other systems. Its principle is the oxidation of ferrous to ferric ions by radiation, the reaction taking place in aerated 0·8 N sulphuric acid; the ferric ion formed is determined colorimetrically with thiocyanate. Its useful range is 4–40 krad; the lower limit is set by the necessity of getting enough ferric ions to measure accurately, and the upper limit by depletion of dissolved oxygen. The response of the system is measured by the radiation yield or G value, the absolute number of ferric ions produced per 100 eV. It depends on the quality of the radiation (or the LET), though not on the dose rate: typical G values are

12–15 for beta and x-radiation, and 15–17 for gamma. In terms of concentration, this is about 16 μmol per l per 1000 rad. Statements on measured doses of radiation should always be accompanied by a note of the *G* value. Great care is needed to get reliable results, and reagents and glassware must be scrupulously clean; the addition of chloride ion or ammonium ion suppresses side reactions and impurity effects, and thallous sulphate prevents undesirable oxidation by organic peroxides. The sensitivity may be extended into the 0–100 rad range by using iron labelled with ^{59}Fe, the $Fe(CNS)_3$ complex being extracted with iso-amyl alcohol and counted. (This use of radiation to measure radiation seems a particularly ingenious application, on the principle of 'set a thief to catch a thief'!) Addition of 0·001 mol/l of Cu^{++} reduces the oxidation rate of the Fe^{++} and increases the range to 80 000 rad; increasing the concentration of Fe^{++} and bubbling oxygen through the solution raises it up to one Mrad (10^6 rad). The dosimeter has also been extended to the measurement of neutron fluxes of 5×10^4 neutrons/cm², by adding benzoic acid to the system (which increases sensitivity) and using iron-59 as label. Boric acid, H_3BO_3, is also added to help trap the neutrons.

6.4.8 Colorimetric dosimetry

Colorimetric methods of dosimetry use the increase in optical density which certain dyes undergo when exposed to high doses of radiation. Orton (1966) describes dosimetry with Perspex incorporating a red dye: this is a commercially available product, and the chemical nature of the dye-stuff is not stated. The dose–response curve is linear from 0·2 to 1·5 Mrad. Clear Perspex also increases in optical density with a linear response between 0·06 and 3 Mrad (Whittaker and Lowe, 1967).

6.4.9 Other methods

Semiconductor detectors (Section 3.6) are usable in dosimetry; e.g. small silicon diodes have been applied to the dosimetry of proton beams (Koehler, 1967; IAEA, 1967).

Other suggestions are: free radical yields in polytetrafluoroethylene (Judeikis, 1968); quinine sulphate (Rytilä and Spring, 1969); thermovoltaic dosimetry (Toole, Henisch, and Miyashita, 1967); and silver nitrate (Law, 1969).

6.5 RADIATION IN PERSPECTIVE

6.5.1 Assessment of radiation doses (MRC, 1960)

The strictness of radiation protection standards is decided by a consideration of the *greatest risk* that can be tolerated. This is the induction of long-term and delayed effects (Section 7.4.4), which are the greatest hazards of radiation, since they are cumulative, may be genetic, require only low doses, and have no threshold (Section 7.4.1). This last point means that any exposure to radiation, however small, may be harmful, so that there is no simple answer to the question of how much radiation is 'safe'. However, with very low doses the risk of serious injury is extremely small, and exposure to radiation is acceptable provided that this risk remains negligible; the object of radiation protection is to minimise it. There must thus be a compromise between the benefits gained from the use of radiation and the hazards of the increased exposure entailed therein; the need to avoid unacceptable exposure to radiation limiting the extent of its use.

The question at once arises as to what is meant by 'very low doses', and the key to the answer lies in the assessment of man-made radiation against the natural background radiation which everyone receives. Background can and undoubtedly does have some effect on humans, but it is difficult, if not impossible, to say exactly how harmful it is; in any case, the question is somewhat academic, for no-one can escape background, the potassium-40 in the body, if nothing else, seeing to that. It seems reasonable to suppose that man is now resistant to natural background, having lived with it long enough for any very radiosensitive strains or individuals to have been eliminated. Within the limits of present knowledge, extra radiation from man-made sources may be defined as 'safe', and the risk from it as negligible, if it does not add significantly to the already-present natural background, or is the same as that received in moving from a low- to a high-background region. It is perhaps fortunate that radiation does occur naturally, otherwise the effects of the man-made variety might well have been much more serious.

6.5.2 Natural background radiation

This radiation has two components: cosmic radiation and terrestrial radiation. *Cosmic radiation* is primarily protons, with smaller proportions of other heavy nuclei; their average energy is about 6000

MeV. They react with the atoms of the atmosphere to produce secondary cosmic rays, of which neutrons, mesons, and gamma radiation reach the surface of the earth. The soft tissue dose is about 30 mrad/year at sea level, of which about 0·3 to 1·1 mrad/year is from neutrons; it doubles for each mile above sea level, for the first few miles. High-flying aircraft may encounter relatively intense cosmic radiation, and a system for automatic warning of hazardous intensities has been developed for the Concorde airliner. Astronauts are deprived of the protective effect of the atmosphere against cosmic radiation, and they must be adequately protected if they are to spend any great length of time on the Moon's surface or elsewhere in space.

Terrestrial radiation sources are found both within and without the bodies of organisms. *External* sources give about 50 mrad/year of gamma radiation, depending on the rocks and minerals in the locality: the dose from granite is higher than from limestone or sandstone, since it contains potassium-40 in the feldspar, and from uraniferous or thorium-bearing rocks the dose is, of course, higher still. The principal *internal* radiation source is potassium-40, which gives about 15–20 mrad/year; the dose from carbon-14 is about 0·7–1·6 mrad/year, from radium and polonium in bone about 0–3, and from radon in the lungs about 0·3. In round figures, the total background radiation averages about 100 mrad/year; there is a 20% variation between different parts of Great Britain, and a 400% variation over the whole world.

Background has not remained constant since the beginning of the world, for there must obviously have been more primordial isotopes in existence then than there are now. Cosmic radiation has probably varied from time to time; some results of radiocarbon dating (Section 5.6) are consistently in error, which indicates that the specific activity of atmospheric $^{14}CO_2$ has changed from its present value. This could be accounted for by the rate of carbon-14 formation having been changed by a change in cosmic radiation intensity. On a longer time scale, there have been occasional fluctuations in the steady progress of evolution, which could have resulted from variations in cosmic radiation altering mutation rates.

6.5.3 Man-made radiation

This radiation arises chiefly from medical exposure and from fall-out (Section 8.1), with some other minor or localised sources.

Medical radiation

Medical radiation is extremely variable. Almost everyone at one time or another receives a diagnostic chest x-ray, but only comparatively few undergo radiation therapy, so that while diagnostic doses can be expressed with some meaning as an average over the whole population and included in overall estimates of radiation exposure (e.g. Penfil and Brown, 1968), this is not possible for therapeutic exposure. The exposure from diagnostic radiation is taken to average about 30 mrad/year, which is perhaps on the high side. The irradiation is usually more or less localised, and the gonadal dose may be very much less than that to other sensitive tissues, as in dental x-rays. A chest x-ray delivers between 0·1 and 6 mrad to the gonads of a man, and between 0·1 and 15 in a woman; the gonadal dose from a pelvic x-ray is about 1 rad. Unfortunately, radiation sometimes appears to be used clinically merely as a routine, with no thought as to whether it is really essential, rather than as a last resort when all else has failed. Sometimes, too, there appears to be too little comprehension of the radiation dose given to the patient, and there are some alarming cases where x-ray exposure has been 50 times greater than necessary. Clearly there is a continuing need for education in the hazards of radiation.

Fall-out

The estimation of average exposure from fall-out is little better than inspired guesswork, because of the varied pattern of deposition over the world. In 1963, the year of highest total fall-out doses, estimates of exposure in the United Kingdom were: external, 3 mrad; internal caesium-137, less than 5 mrad; strontium in bones of young children, 23 mrad, and in bone marrow, 10 mrad. Fall-out exposure will continue for a long time for, as the reservoir of activity in the atmosphere falls out, it enters the soil or the sea, forming another reservoir from which it may be taken up by plants and animals and enter human diet. Besides being expressed as yearly dose rates, exposures from fall-out are therefore given also as the *dose commitment*, the total dose that will be delivered in the future from releases of activity that have already taken place. The estimated dose commitment from 1954 to the year 2000, arising from weapons testing up to 1962, is shown in Table 6.1. The total dose commitment to the gonads is about 75 mrad, which would be received from natural background in about 9 months;

Table 6.1. ESTIMATED DOSE COMMITMENT, FROM 1954 TO 2000, FROM VARIOUS SOURCES

Source	*Dose Commitment* (mrad)
External short-lived isotopes	20–25
External caesium-137	25–30
Internal caesium-137	15
Internal carbon-14	13–20
Internal strontium-90 on bone surface	150
Internal strontium-90 in bone marrow	80

for the cells lining bone surfaces it is about 250 mrad, or 30 months' background, and for bone marrow about 160 mrad, or 20 months' background. In terms of an annual exposure, the dose commitment is reckoned to be equivalent to about 8 mrad/year, though the actual annual doses are high at first and then fall off. Carbon-14, by reason of its long half-life compared with other fission products, will continue to deliver low doses for a prolonged period of time after the year 2000, and eventually the total dose from it will be as great as, if not greater than, that from other nuclides, though it will be spread over several generations.

The dose commitment from fall-out should not increase provided that there are no more atomic weapon tests in the atmosphere, or (which Heaven forbid) a nuclear war. The nuclear test ban treaty is not binding on countries that have not signed it, and in view of past experiences there is room for doubt as to whether pieces of paper are completely effective safeguards to prevent people doing what they should not. Besides fall-out, it would also be wise for a watchful eye to be kept on releases of activity into the environment from peaceful uses of atomic energy, which will increase as time goes on.

Other Sources

Other sources are very minor, totalling not more than 2 mrad/year. They include such 'appurtenances of civilisation' as *luminous clocks and watches;* the radium formerly used has largely been replaced by tritium and other low-energy beta emitters, so that there is likely to be far less hazard from newer timepieces than from older ones. At this point, the luminous discs once used by the Armed Forces, e.g. for map reading, may be mentioned; even now, thirty years after the 1939 war, such a disc can still churn out 1 mrad per hour at 3 in (75 mm) distance. *X-ray fluoroscopy* was once a very popular 'gimmick' for shoe-fitting, though it was never really essential to secure a good fit; the author has personal recollection of seeing his foot bones on the screen in childhood days, but has not come across such a machine for years. As fashions come and go, so fluoroscopes seem largely to have gone.

Television receivers have an x-ray output of about 1 μrad per h per mA; as this is critically dependent on the working kilovoltage of the tube, colour sets deliver more than the black-and-white. The dose to the viewer is considerably lessened by the glass cover in front of the tube and by the usual viewing distance from the set, and on average it is infinitesimal compared with background, at about 1 μrad per year. The ICRP maximum limit is 0·5 mrad/h at 50 mm from the tube face.

Occupational exposure

All the above sources of radiation affect everyone, and this non-occupational exposure is distinguished from the occupational exposure of people who are in contact with radiation or who handle radiation sources during the course of their daily work; this is a localised occupational hazard that does not affect the general public. Its sources include nuclear reactors, which present a negligible risk because of their built-in safety, and industrial and research uses of radiation. Safety precautions and routines limit exposure to a minimum, and the rare accident is almost always the result of non-observance of these. The estimated gonadal dose to radiation workers is 0·2–0·4 mrad/year. Laughlin, Vacirca, and Duplissey (1968) quote a somewhat bizarre source of exposure, the handling by embalmers and physicians of corpses of patients who were treated internally with radio-isotopes (Section 5.4) shortly before death; with an isotope like iridium-192, of relatively long half-life, there is the possibility of appreciable radioactivity remaining in the body at death.

Gonadal doses

The average gonadal doses, i.e. the genetically significant radiation doses, from man-made sources may now be summarised (in Table 6.2). On average, these doses are much less than the background of 100 mrad/year, so that the major proportion of the total expo-

Table 6.2. APPROXIMATE AVERAGE YEARLY GONADAL DOSES FROM MAN-MADE SOURCES

Source	*Dose* (mrad/year)
Fall-out	7
Diagnostic radiology	30 (very variable)
Occupational	0·3
Other	≤ 2

sure arises from natural sources; but it must be remembered that an average includes both high and low values, and there may well be particular areas or groups of people whose exposures are considerably higher. The doses to the gonads from artificial

Table 6.3. APPROXIMATE GONADAL DOSES FROM ARTIFICIAL RADIATION SOURCES, AS PERCENTAGES OF THE NATURAL BACKGROUND

Source	*Dose* (%)
Fall-out	8
Diagnostic radiology	14–20 (very variable)
Occupational exposure: Radiology	1·6
Occupational exposure: Atomic energy	0·1
'Appurtenances of civilisation'	$\ll 1$

radiation sources, as percentages of the natural background, are shown in Table 6.3. The average exposure of the gonads to radiation in the first 30 years of life is estimated to be about 8 rad, made up as shown in Table 6.4.

Table 6.4. APPROXIMATE AVERAGE EXPOSURES OF THE GONADS TO RADIATION FROM VARIOUS SOURCES IN THE FIRST 30 YEARS OF LIFE

Source	*Dose* (rad)
Natural background	
Cosmic	1
Terrestrial	1·5
Atmospheric	0·1
Internal	1
Total	3–4·5
Man-made sources	
Fall-out	0·1
Diagnostic	$\ll 3$ (very variable)

6.5.4 Maximum permissible doses (MPD) (ICRP 1966)

These doses are based on the risk of delayed radiation effects (Section 7.4.4), and on the doses to critical organs (Section 6.1.4) and *critical populations*, the group of people who are at greatest risk of radiation injury in given circumstances.

In *occupational exposure*, no-one under 18 may work with radiation; after this age, a total of 5 rem/year of whole-body radiation is permitted. If N is the age in years, the maximum permissible accumulated dose is $5(N—18)$ rem, subject to a maximum of 50 rem in the first 30 years up to age 48. Beyond this age, the possibility of parenthood recedes rapidly, so that 50 rem are allowed in the next 10 years and 200 rem in the whole lifetime. However, anyone starting radiation work comparatively late in life should still keep within the 5 rem/year limit. Maximum whole-body doses allowed for shorter periods are as follows:

13 weeks (3 months),	3 rem
1 week	0·3 rem
1 day	0·1 rem
1 hour	12·5 mrem

The extremities (hands and feet) may receive more than the whole body, up to 1·5 rem/week, 20 rem/quarter, and 75 rem/year. Rather more hard beta is allowed than x- and gamma-radiation, since it is less penetrating and has less effect on the haematopoietic tissues and gonads.

The comparatively higher doses for shorter times are only allowed provided that the other figures are not exceeded, since a high dose rate is permissible only for a short time and, if exposure continues, it must be at a reduced rate. Thus, if the whole of the weekly allowance is used up within a day (as may happen in clearing up after an accident), no more radiation work whatever is permitted for the rest of that week, so that the permitted dose may, as it were, catch up with the dose received. Occasionally it may be necessary to exceed the MPD, because there is no other way of doing a particular job. This is the only justification for a *planned special exposure*; it must be worked out beforehand, and cannot be applied to an accidental over-exposure afterwards. The limit is four times the quarterly dose, i.e. 12 rem, and an individual may receive planned special exposures only infrequently.

The weekly allowance is taken in practice as the rule-of-thumb working guide, though it should be reduced to 0·1 rem/week if it is to agree with the yearly figure of 5 rem.

Particular care must be taken with young people—hence the above age limit of 18—and with women of child-bearing age, who are not allowed planned special exposures, and for whom the MPD at the abdomen is reduced to 1·3 rem/quarter, or 2·5 mrem/hour at constant exposure rate over the 40-hour working week. Embryos in utero are the critical population for whole-body irradiation so that, during pregnancy, the allowances are cut to 150 mrem/month, or 1 mrem/hour at constant exposure rate, and the wearing of an extra film badge at the waist is recommended. For the same reason, medical radiation of women should be restricted to the fortnight immediately following menstruation (Section 7.4.5).

Designated radiation workers have regular health checks and continuous recording of their radiation exposure (Section 6.6.10), so that the MPDs for occupational exposure may be set higher than for the general public, who are not so supervised. *Non-occupational allowances* are $\frac{1}{10}$ of the above figures, with a limit of 5 rem total cumulative dose in the whole lifetime.

The long-term effects of radiation are the most serious, but they take a long time to show, and it is useful to have an *early indicator* of biological radiation damage. Designated workers have an annual blood count, though haematological changes appear only after comparatively large doses (Section 7.4.3); chromosome damage is more sensitive (Section 7.3.4); and reduction in motility of sperm (Section 7.4.3) even more so. Other indicators are: plasma β-amino-iso-butyric acid (Section 7.3.3); changes in the fingerprint pattern warn of long-term skin damage (Section 7.4.3); and decrease in plasma volume and the NVD syndrome indicate severe injury (Section 7.4.3).

Since delayed effects of radiation are cumulative, exposure should be kept to a minimum, and MPDs are, of course, upper limits; they have, in fact, been reduced from the values laid down when radiation safety standards were first introduced. However, the term 'permissible' carries some implication of permission to have a certain dose, from which it is not far to the assumption of a right to have that dose. It is conceivable therefore that someone receiving a small radiation dose could 'stand up for his rights' and demand more, so as to bring him up to the permitted allowance, though admittedly nobody with any sense would do so. For this reason the term 'maximum acceptable dose' has been proposed as being preferable to 'maximum permissible dose', as it removes this implication.

6.6 RADIOLOGICAL SAFETY—THE VITAL NECESSITY

The importance of safety in radiation work is paramount, and can hardly be over-emphasised; it must be designed and built in right from the start. The reason for this is made clear in Section 6.1.2, and radio-isotopes and ionising radiations are handled with the same respect as poisons, dangerous drugs, or pathogenic micro-organisms. At the same time, there is no need to exaggerate the risks, or to panic whenever the dreaded word 'radiation' is mentioned: in many tracer experiments, no more is required than the discipline and cleanliness of a high-quality chemistry laboratory and, provided that proper precautions are observed, there is less hazard from radio-isotopes than from many common chemicals that are corrosive or explosive. But, as with these latter, carelessness with radiation can lead to real dangers, and the reasoning behind the recommended safety precautions must be clearly understood. For the same reason, special regulations govern the transport of radio-isotopes.

Legal requirements, contained in the Radioactive Substances Act 1960, Ionising Radiations (Unsealed Radioactive Substances) Regulations 1968, Ionising Radiations (Sealed Sources) Regulations 1969, and others, govern the holding and use of radiation sources; all holders and users must be licensed. Certain conditional exemptions apply to the use of isotopes in schools for teaching, and to natural uranium and thorium compounds; one is also allowed to keep a luminous watch or clock! The requirements concerning radiological safety are laid down in Codes of Practice for protection of radiation workers; an 'official' code is published (Ministry of Labour, 1964), and particular establishments may have their own local codes. Boursnell (1958) gives an excellent coverage of the safety techniques that should be observed with radio-isotopes.

All laboratories handling radiation sources have a set of Safety Rules. They vary slightly in detail from one place to another, but all follow the same general pattern. They are based on the principles of safety in working, containment, or restricting the spread of activity, and responsibility to others, for it is the duty of anyone who works with ionising radiations to avoid exposing other people to a dangerous situation of which they may be unaware.

The precautions to be observed may seem at first sight to be very elaborate; guarding against things that would rarely, if ever, happen. But the nature of radiation is such that things must be guarded against; there is no room to chance it and hope for the best, and work must be organised on a fail-safe basis. There are some alarming cases on record where neglect of simple precautions has led to contamination being spread far and wide, farther and wider than might ever have been thought possible.

The safety rules for the isotope laboratory boil down to good laboratory discipline and housekeeping and common sense, which any laboratory worker should have anyway, but which become particularly important with isotopes. The rules are discussed under the following ten headings:

6.6.1. Careful planning;
6.6.2. Erring on the safe side;
6.6.3. Protective clothing;
6.6.4. No mouth operations;
6.6.5. Minimising radiation dose;
6.6.6. Keeping activity in its proper place;
6.6.7. Tidy working;
6.6.8. Waste disposal;
6.6.9. Cleaning up and decontamination;
6.6.10. Health checks.

6.6.1 Careful planning

Careful planning of a radiotracer project is essential, not only so that the best possible results may be obtained (Section 4.4.1), but also to minimise the hazards to the experimenter with respect to both irradiation and contamination. Advice should be sought from the person responsible for the safe running of the laboratory, or at least from a colleague, in order to bring to light any hazards that may have been overlooked; there is nothing whatever to be ashamed of in asking for advice in this way. In many laboratories, a system operates whereby a project has to be submitted for approval before work can start, thus providing an independent assessment of its safety.

Careful planning will avoid any foreseeable dangers, but some cannot be foreseen and, for this reason, a *trial* or *dummy run* is performed, in which everything is carried out in exactly the same way as the experiment proper except that the isotope is not introduced. The dummy run is of great value to avoid getting into a radioactive mess, and should never be omitted when a new experimental technique is introduced. If an unforeseen hazard is revealed, the experiment can be modified before the isotope is introduced.

6.6.2 Erring on the safe side

This means being over- rather than under-cautious, and exemplifies the fail-safe principle. Thus, all estimates when calculating doses of radiation are made pessimistic, so that, if a mistake should be made, the hazard is not increased. Sources are assumed to be at full strength, any previous use or decay being ignored; and all their radiation is assumed to be directed at the user, under broad-beam conditions (Section 2.5.2). Apparatus is assumed to be contaminated, and treated accordingly, unless it is definitely known without any doubt to be free from radio-isotopes. Detailed instructions on what to do in the case of accident are laid down, and they must be known and followed.

6.6.3 Protective clothing

Such clothing prevents contamination of the skin and of the ordinary garments underneath, both of which are undesirable. Radioactivity on the skin not only presents an external radiation hazard, but may also enter the body. Many chemicals can penetrate intact skin, and will get in if the skin is broken by any kind of wound. All injuries, even if only very minor, must be covered, especially those on hands and forearms.

Protective clothing takes various forms, depending on the amount of activity being handled; for tracer work the ordinary lab coat and rubber, polythene, or similar gloves are usually adequate. Overshoes, respirators, face-masks, and other more elaborate means of protection are used in more dangerous situations. A point to note is that, if a waterproof apron is worn to guard against splashes, liquid can run down it on to the legs and feet unless precautions are taken. Disposable paper lab coats have been developed recently and are more satisfactory in that they eliminate the need for special laundering arrangements, for they can simply be thrown away when soiled or contaminated.

Protective garments should be put on and taken off in a changing area, away from the laboratory proper. Coats, jackets, books, etc., are left in this area during work, and the protective clothing is left there afterwards. It may be noted that protective clothing has to be put on and taken off in such a way that the outside, possibly contaminated surfaces never come into contact with what they are protecting. The same routine is followed even with new clothing, to avoid developing bad habits.

Rubber gloves often seem to give trouble in this respect; the procedure for putting them on is as follows:

1. Dust the dry hands with a coating of talcum powder.
2. Fold back about 2 in (50 mm) of the cuff of each glove; do not touch the outside of the glove with the bare fingers.
3. Put on one glove, gripping only the exposed inside of the cuff with the other hand.
4. Put on the other glove, inserting the gloved fingers of the first hand into the fold, so that they touch only the outside of the glove.
5. Unfold the cuffs, ensuring that the gloved fingers do not touch the inside of the gloves.

To remove gloves (having washed and monitored them) the procedure is as follows:

1. Grasp a fold of the cuff of one glove with the other hand, and peel off the glove—pulling it off by the finger ends is liable to split it.
2. Insert the fingers of the now-ungloved hand inside the cuff of the other, grasp it, and peel it off.
3. The gloves are now inside out; when they are dry (they inevitably become damp with perspiration from the hands), turn them right way out, taking care not to touch the outside with the bare fingers, and leave a fold at the cuff ready for the next time they are used.

Wearing gloves for a long time can become uncomfortable and, as it is not necessary to keep them on at all times, the experiment can be planned so that they are worn for a minimum of time, i.e. only when it is essential. It may sometimes be useful to have one hand gloved and the other ungloved. Gloves are left on paper on a special tray when not in use.

6.6.4 No mouth operations

Mouth operations are forbidden, to prevent any activity entering the body by this route; in particular, pipetting carries an ever-present risk that liquid may be sucked up accidentally. This is of no consequence with many solutions in an ordinary laboratory, but though the chance of an accident is very slight, if care is taken, it is quite unacceptable with a solution that is even slightly radioactive. Indeed, mouth pipetting of any solution in an isotope laboratory cannot be countenanced, for the pipette might pick up contamination from the bench and transfer it to the mouth next time it is used; and one can make a mistake and pipette the wrong solution. One of the many types of 'safety' pipette is used instead of the mouth.

Glass blowing must never be carried out on apparatus from the isotope laboratory, for it is a mouth operation, and also activity adsorbed onto the glass may be vaporised on heating, giving rise to airborne contamination.

Eating, drinking, smoking, and the application of cosmetics are also strictly forbidden in any radiation area.

6.6.5 Minimising radiation dose

Apart from its strength, three key factors govern the radiation dose from a source: the time and distance of exposure, and how the source is shielded.

Strength of the source

The desirability of using no more isotope than is necessary in an experiment has been mentioned above (Section 4.4.4), though often a comparatively large source has to be obtained and handled, the small amount for the experiment being dispensed from it, and this is perhaps where the chief danger lies. Large sources can deliver a fair dose of radiation quite quickly, and they should therefore be exposed and manipulated for as short a time as possible, and when not required stored with adequate shielding at a distance from the working area. However, a properly conducted tracer experiment may easily give less dose than a luminous watch with radium on it.

Distance

Distance is of great help in reducing the radiation dose from a source, since the inverse-square law operates, and in fact with most tracer sources it may be the only precaution required. For example, a phial of 1 mCi of ^{131}I solution at 3 mm distance, as it might well be when dispensing, will give a week's permissible dose (300 mrem) in some 40 s whereas it gives only 100 mrem in 45 h at 1 ft (300 mm) away, so that it could safely be left at this distance for the whole working week. At the other side of the room, the dose is hardly distinguishable from background. The same reasoning applies in a major accident with a source—absence of body is better than presence of mind—so one gets out of the area as quickly as possible (remembering to extinguish heaters before leaving, to prevent outbreak of fire), and thinks what to do afterwards, at a safe distance. There is no point in being heroic if it results in radiation injury, unless it is to rescue someone else from danger.

Sources should never be picked up in the fingers and, even with gloves on, tongs or forceps should be used. Gloves will prevent contamination but will not stop gamma and strong beta radiation reaching the hands. At the same time, there is no need for over-elaborate handling precautions, for remote handling may be so clumsy and slow that more direct means of performing the same operation will be less hazardous, quicker, and easier. A source should never be looked at closely, as the eyes are sensitive to radiation.

Desk work should be done in a separate office rather than in the laboratory where the activity is handled.

Shielding: from alpha and beta

Shielding in general (Jaeger *et al.*, 1968) is best considered on a basis of how the radiation interacts with matter, and becomes more difficult as the radiation increases in penetrating power. Thus, as mentioned earlier, alpha particles present no difficulty whatever, being stopped by a few inches of air or a thin layer of anything else. Beta radiation is likewise easy to deal with, as the container is enough to provide a satisfactory shield except for large sources of hard radiation, which are better shielded by low atomic number materials (e.g. water or concrete) than by lead, or the bremsstrahlung produced (Section 2.4.3) may be more of a problem than the beta itself, the shield aggravating the radiation hazard rather than reducing it.

Shielding: from x- and gamma-radiation

By contrast with alpha and beta, the trouble with x- and gamma-radiation is that it cannot be completely stopped, but only attenuated, and shielding becomes of great importance, particularly with high-energy x-ray sets and with large gamma sources. All-round shielding is needed, not forgetting floor and ceiling, to cut down scatter of the radiation and also to protect people in neighbouring rooms, for other people's gammas are deservedly unpopular! Lead or concrete provide good protection. *Half-thicknesses* are useful in calculating shielding requirements for gamma (Section 2.5.2), the number of half-thicknesses (N) needed to reduce the radiation intensity by a desired factor (X) being related simply by the equation: $N = 3{\cdot}33 \log_{10} X$. Thus 3 half-thicknesses are needed to reduce the radiation to 10% of its incident intensity, and about 7 to reduce it to 1%. Specimen figures for half-thickness are given in the first column of Table 6.5, which shows the shielding

Table 6.5. SHIELDING NEEDED FOR 1 MeV GAMMA-RADIATION

Material	*Thickness* (in) *for % transmission of:* 50%	10%	1%
Lead	0·35	1·5	2·75
Iron	0·6	3·3	6
Concrete, water	4	12	23

for 1 MeV radiation. The UKAEA has developed uranium as a shielding material for gamma radiation; the 'depleted' element, uranium-238 from which the uranium-235 has been removed, is used, and is marketed as 'industrial uranium'. It is lighter and more compact than the equivalent amount of lead.

Shielding: of x-ray sets and electron microscopes (NCRP, 1968)

In the shielding of x-ray sets and similar machines such as electron microscopes, the main emphasis is on enclosure and adequate shielding of the radiation source. This is to cut out narrow stray beams and localised scatter, in whose presence there is unlikely to be much, if any, correlation between radiation doses received at one part of the body and those indicated by a dosimeter worn at another part. The shielding should fulfil the following conditions:

1. No-one should be able to expose inadvertently any part of himself or of anyone else to any primary beam of radiation, whether or not it is collimated.
2. Exposure to diffracted beams and scattered radiation must be prevented, so far as is practicable.

3. Everyone nearby must have adequate warning that the x-ray tube is energised.
4. The methods for achieving these three conditions must be reliable, proof against tampering, continuously effective, and wherever practicable twofold, so that two independent safety devices must fail before there is any possibility of inadvertent exposure to a primary beam.

The shielding of x-ray sets should cut down radiation levels to a maximum of 0·75 mrem/h for non-designated workers, or to 2·5 mrem/h for designated workers.

Shielding: from neutrons

Neutrons are the most difficult radiation to shield, because of their great penetrating power and their energetic reaction deep within the irradiated material (Section 2.4.4). Materials containing much hydrogen, e.g. water, polythene, or paraffin wax, are used to stop them.

Shielding: of nuclear reactors

Shielding nuclear reactors presents quite a problem to the health physicist, for both gamma radiation and neutrons have to be dealt with. The first layer of the shield (nearest the source) must be able both to moderate fast neutrons and to attenuate gamma. When the neutrons have been degraded to thermal energies, an element of high neutron capture cross-section must be provided. Finally, when the neutrons have been removed, an efficient gamma shield is needed. No one material can do all these jobs, so that a multi-wall shield must be built, its design being dictated by careful consideration of the properties of the radiation flux.

A recent development (Hall, 1966) that has some potential in this field is metallic concrete, which is commercially available. This is like ordinary concrete except that it contains a high percentage of lead or other metal, the composition being adjusted to suit the particular requirement, so that it combines good shielding qualities with ease of working and forming.

6.6.6 Keeping activity in its proper place

This is essential to avoid unnecessary risks to health, and several lines of defence are erected to do this. If an accident occurs, and the first line fails, others come into operation automatically, so that the situation is kept under control and does not suddenly turn into a major crisis. Thus operations with liquids are carried out in double vessels so that, if a breakage occurs, the activity is contained within the outer one and does not spread; and work is done on impervious plastic or metal trays lined with disposable absorbent paper, which restricts the spread of activity should a liquid be spilt. Bench tops and floors should also be impervious, for if contamination should soak into them it is expensive and difficult to remove: there are cases in which the whole of an excessively contaminated laboratory had to be removed as radioactive waste!

Radiation work should be *segregated* from 'non-active' work, and the two should not be mixed in the same laboratory. Glassware can (quite cheaply) be '*indelibadged*'—a label with the international radiation warning trefoil symbol is stuck on and permanently fixed by heating—so that it is immediately recognisable if it gets anywhere it should not, e.g. to the glass-blower.

Dry and dusty materials are handled in glove boxes or fume cupboards, to prevent air contamination and, for the same reason, heating is best carried out in the fume cupboard. Evaporation of a liquid can cause fine spray and invisible air contamination, and is best done on a water bath or under an infra-red lamp, again preferably in the fume cupboard.

High-activity sources are confined to a 'hot' area, labelled accordingly and in an unfrequented part of the laboratory, remote from the entrance. Dispensing is carried out in the 'hot' area, and the sources are not taken away for this purpose. They are, as mentioned above, kept shielded when not in use.

The final line of defence, after the double vessel, the tray with its paper lining, and the impervious bench, is the exit from the laboratory. Anything leaving an isotope laboratory should be monitored for freedom from contamination, and in fact this is now a legal requirement for laboratories handling unsealed sources. The outer door must bear a metal plate with the international radiation warning trefoil symbol embossed on it, so that, even if the paint should be burnt off in the event of a fire, the warning will still be visible (Figure 6.3). A monitor at the door can give the alarm if anything radioactive is inadvertently removed from the laboratory (Kemp, 1969).

It is important not to spread contamination with radio-isotopes to objects that cannot easily be decontaminated or that may be touched unknowingly by someone else before decontamination: objects such as electricity switches, water taps, reagent bottles, etc., that are in common use and not restricted solely to one worker. A glove could transfer activity to them, and it could then be picked

Figure 6.3. Radiation warning door plate: painted vividly in yellow, purple, or red

up later by an ungloved hand: such objects are therefore handled by means of paper tissues that are thrown away after use.

It is also important, especially with low-activity work, to avoid contamination of the laboratory around and on the counting gear, not only from the health hazard aspect, but also because a high background count rate may make accurate work with small sources impossible. For this reason the counter is never operated with gloves on, and the only activity allowed in the counting area should be the actual samples for counting: thus, if a spillage should take place, the amount of activity released is minimal. Apart from this, the mere presence of a large source can increase background significantly in high-precision work.

It is here implied that the counters are set apart from the rest of the laboratory, and that samples for counting must therefore be carried from the bench to the counting room. This is an obvious weak link in the chain of safe working, and attention must be paid to it in designing the laboratory layout. If the laboratory is a conversion of an existing one, rather than the ideal 'purpose-built' suite, it may indeed be found preferable to have the counter alongside the bench, contamination and increased background being rated less serious than the hazard occasioned by carrying samples around the laboratory.

6.6.7 Tidy working

Tidy working is one of the cardinal virtues in handling isotopes, for a safe laboratory is a tidy one. Nothing that is not needed for the experiment is allowed near the working bench, particularly books, etc., that would be difficult to replace if contaminated: if they are not there, they cannot become contaminated. Apparatus is arranged on the bench so that it can be reached easily and is not liable to be knocked over, and work is carried out unhurriedly, avoiding rush and sudden movements that could cause an accident.

UKAEA RADIOACTIVE MATERIAL ... RADIO

Figure 6.4. 'Radioactive tape': brightly coloured in yellow, purple, or red

A complicated sequence of operations is mapped out beforehand. An essential part of tidy work is knowing the identity of everything, radioactive and inactive alike; containers should be labelled clearly, preferably before anything is put into them, with a statement of the contents. Note that stick-on labels must be moistened by means other than the tongue (Section 6.6.4). Printed self-adhesive tape, the so-called radioactive tape, may be had in different designs (Figure 6.4), and is very handy, apart from the disadvantages attached to all self-adhesive labels. (It turns up surprisingly often adorning student laboratory notebooks on which no trace of radioactivity can be detected!) Small 'tie-on' labels secured with rubber bands may be used when there is no risk of the label being caught and pulling the container over.

6.6.8 Waste disposal

Great care has to be exercised in the disposal of radioactive waste, for it is perhaps the most hazardous operation of any, in that it exposes the general public to radiation and to radio-isotopes without their being aware of it, and also, once activity has been released, it is impossible to recover it or control its fate. Disposal is therefore strictly controlled by law, and in fact radioactivity is one of the few wastes whose disposal is internationally controlled.

Waste diposal is based on two principles: dilution and dispersal, or concentration and containment.

Dilution and dispersal

Dilution and dispersal recognises that a release of activity into the environment will not be hazardous as long as it does not significantly increase the already-present natural background, or introduce large quantities of isotopes that are particularly dangerous, such as alpha-emitters. If a comparatively small amount of isotope is dispersed throughout a large bulk of inactive carrier material, in the same chemical form, the specific activity is reduced virtually to vanishing point, and the isotope may then be released without harm. This principle is most applicable to isotopes of elements that are abundant naturally, which is often so in biological work.

The Ministry of Housing and Local Government therefore gives permission for small quantities of all but a few isotopes to be disposed of in the ordinary inactive solid waste, through the drainage system, or by burning. The amounts permitted depend on a number of factors, such as the total water consumption of the establishment. They are a total maximum; each individual worker is restricted to a small fraction of this, and he must make sure he knows what his allowance is.

Specimen figures for disposal of activity (taken from the authorisation of the University of Aston in Birmingham) are as follows.

Solid waste may contain not more than 20 $\mu Ci/ft^3$ of carbon-14+tritium, and not more than 2 $\mu Ci/ft^3$ of all other radionuclides excluding decay products, the maximum allowable in any one solid article not to exceed half of these figures.

Liquid waste may contain up to 100 mCi of carbon-14+tritium per calendar month, and 5 mCi of all other radionuclides; alpha emitters are not permitted.

Burning is allowed of not more than 100 mCi of carbon-14+tritium, 10 μCi of alpha emitters, and 1 mCi of all other radionuclides, per calendar month. The gas or vapour from the burning must not be able to enter a building so far as is reasonably practicable.

The waste should be mixed thoroughly with carrier before disposal, especially liquid waste—it is not sufficient to pour the active liquid away and wash it down with carrier afterwards, for the activity may be trapped and held stagnant at certain places in the plumbing, or be selectively taken up by micro-organisms. Likewise, rubber mats should not be placed in the sink, as they can obstruct the free flow of liquid into the drain.

Concentration and containment

The principle of concentration and containment operates with large quantities and/or hazardous isotopes, whose release would be dangerous and thus cannot be allowed. The source is stored out of contact with the biosphere for long enough to allow its activity to decay away to the point where it will not be hazardous if it is then released. The principle is seen on the small scale in the laboratory, with short-lived isotopes of half-lives up to a few weeks (e.g. phosphorus-32): they are simply stored in a safe place until the activity has gone, a useful practical guide being that the activity falls to 1/1000 of its original value in 10 half-lives.

A far greater problem arises with, for example, nuclear reactor installations, where activities are in tens or hundreds of curies. The waste is concentrated to occupy as small a volume as possible, customarily by controlled evaporation or incineration, though the concentration of activity from a large volume of dilute liquid waste by adsorption onto a precipitate (Section 4.2.2) or by means of micro-organisms is another possibility. After encasing in concrete, the waste is buried deeply, e.g. in an abandoned mineshaft, or sunk in the ocean deeps, in the hope that it will be safe there for many thousands of years.

6.6.9 Cleaning up and decontamination

An experiment is not finished the moment the last reading has been taken, in spite of the temptation to work out the results at once and leave the clearing up till later—a bench left just as it is at the end of an experiment constitutes a hazard. Safety precautions must not be relaxed at this stage of the experiment, and a definite procedure should be followed. It is not good practice to leave all the clearing up to a technician, unless he has been assisting with the experiment and knows how active everything is; a careful worker will clear up after the experiment himself to a point where a technician can take over without being exposed to hazards of which he is unaware.

Glassware used during the experiment is not allowed to dry, for activity that has dried on to a surface becomes firmly adsorbed and is then much more difficult to shift than if it had been swilled off while still wet. After washing, the glassware is monitored, and not put away until it is free from activity. The ease of cleaning varies greatly according to the compound and isotope being used, and a satisfactory routine should be worked out and adhered to. Powerful detergents, 'specially formulated for the decontamination of radioactive glassware', are available, but they tend to be expensive. Often, too, equally good results may be had simply by rinsing with water, brushing with an ordinary detergent (a biodegradable one, of course), rinsing well again, and allowing to dry; an ultrasonic cleaning machine is an alternative. The special detergents may be reserved for stubborn contamination. A drying oven is not necessary, as any residual contamination that may be left on the glass will tend to become baked on and more difficult to remove. Items should not be left steeping for days in a sink for, even with a tap running, this is not sufficient for thorough cleaning of the inside of a vessel, where the activity usually is. Simple diffusion cannot substitute for proper washing.

The bench should be monitored, and any 'hot spots' that may be found are cleaned off. The gloves are washed and monitored, and finally the hands and body are checked for contamination. Then, and only then, can the experiment be regarded as truly finished.

Monitors for hand and body contamination are currently available that operate automatically for a fixed length of time, to eliminate any subjective error by the user. They give a reading in terms of 'maximum permissible levels' of hand and body contamination, and an indication of 'All clear' or 'Wash again', as the case may be.

6.6.10 Health checks

Anyone handling radiation or working very close to it on a routine basis or over relatively long periods is a *designated* or *classified* radiation worker. He must wear a film badge or pocket dosimeter, and have an annual medical examination and blood count (Section 6.5.4). Sampling for internal isotopes, or whole-body monitoring, may be required in special circumstances. The annual medical and haematological checks provide a 'fringe benefit', in that they may reveal something wrong which has no connection whatever with radiation exposure, and there are cases on record where people have cause to be grateful for such revelations. People in contact with radiation only occasionally, and students under instruction, are not classified and require only a film badge. This has been said to be unnecessary, for the radiation doses received are very small, which is true, but accidents can happen (or more correctly are caused) and the film badge may be the only indication that there has in fact been one. If an accident is known to have occurred, and there is a chance of exposures higher than normal, films should be sent for processing, and dosimeters read, immediately. Apart from this, the wearing of a film badge promotes awareness of the fact that radiation is hazardous. Hughes (1969) discusses the criteria for designation (or not) of people who do not work full-time with radiation.

The issue of a film to designated workers is a legal requirement, not merely a whim of the employer. It affords a certain degree of control of the work being done, and it provides a cumulative record of radiation exposure which is kept at least 30 years and is transferred with the worker if he changes jobs. The wearer of a film badge has certain responsibilities, and must observe precautions in handling it. He must:

1. Wear the badge at all times during the working day;
2. Wear it in the correct position, normally on the lapel of the lab coat, not elsewhere; and the right way round, the yellow side of the film at the back of the badge towards the body;

3. Avoid subjecting it to heat, damp, or chemical fumes;
4. Return it promptly for processing at the expiry date—if it is retained too long, it cannot be interpreted, and is credited with the full MPD for the period (400 mrad/month), i.e. $\frac{1}{12}$ of the yearly MPD.

The wearer must not:

5. Wear anyone else's badge;
6. Use it for other than personal monitoring—badges are issued specially for monitoring apparatus, sources, etc., if this is necessary;
7. Wear it other than in the proper holder;
8. Remove anything from the holder or add anything to it;
9. Mutilate the film in any way, by pressure, bending, or puncturing.

All the radiation that the film records is credited to the wearer, whether or not he is wearing it at the time, and it is therefore important to keep the badge in a radiation-free place when work ceases, and to avoid contamination of the holder. One should be alive to the much greater possibility of the latter if it is necessary to wear the badge on or near the hands when open sources are being used, for an unnoticed droplet of radio-isotope splashing on to it can cause much trouble until it is traced. In circumstances when the hands receive a higher dose than the body, and finger-tip or hand doses are required, dosimeters designed for the job (often the thermoluminescent type) are available, and are preferable to film badges.

REFERENCES

AGLINTSEV, K. K., (1965). *Applied dosimetry*, Iliffe, London, 235 pp.

BATTAERD, H. A. J., and TREGEAR, G. W., (1966). 'Radiation Dosimetry—The Fricke–Miller $FeSO_4$ Dosimeter', *Rev. pure appl. Chem.*, **16**, 83–90.

BECKER, K., (1966). *Photographic Film Dosimetry*, Focal Press, London, 223 pp.

BECKER, K., (1968). 'Recent Progress in Radio-Photo-Luminescence Dosimetry', *Hlth Phys.*, **14**, 17–32.

BECKER, K., (1969). 'TSEE as a Method for Dose Measurement using Lithium Fluoride'. *Hlth Phys.*, **16**, 527–532.

BOURSNELL, J. C., (1958). *Safety Techniques for Radioactive Tracers*, Cambridge University Press, Cambridge, 68 pp.

CHISWELL, W. D., and DANCER, G. H. C., (1969). 'Measurement of Tritium in Exhaled Water as a Means of Estimating Body Burden', *Hlth Phys.*, **17**, 331–334.

DENHAM, D. H., (1969). 'Health Physics Considerations in Processing the Trans Plutonium Elements', *Hlth Phys.*, **16**, 475–487.

EVANS, A. G., (1969). 'New Dose Estimates from Chronic Tritium Exposures', *Hlth Phys.*, **16,** 57–63.

GEIGER, E. L., (1968). 'Tritium Film Badge', *Hlth Phys.*, **14,** 51–55.

HALL, R. M., and WRIGHT, C. N., (1968). 'Comparison of Lithium Fluoride and Film Dosimetry', *Hlth Phys.*, **14,** 37–40.

HALL, W. C., (1966). 'Lead Concrete—First Extra-high Density Shielding Suitable for Installation by Mass-Production Methods', *Nucl. Engng. & Des. (Neth.)*, **3,** 476–477.

HEARD, M. J., and JONES, B. E., (1963). 'A New Film Holder for Personal Dosimetry', *AERE* M 1178.

HUGHES, D., (1969). 'Administrative Criteria for Designation of Workers with Unsealed Sources', *Nature, Lond.*, **224,** 728.

IAEA, (1967). *Solid State and Chemical Radiation Dosimetry in Biology and Medicine*, STI/PUB/21/23, International Atomic Energy Agency, Vienna, 471 pp.

ICRP, (1966). 'Radiation Protection—Recommendations of the ICRP Adopted 17. 9. 65', *ICRP Publ.*, No. 9, Pergamon, Oxford, 27 pp.

ICRU, (1968). 'Radiation Quantities and Units', *ICRU Rep.*, No. 11, Washington, D.C.

JAEGER, R. G., BLIZARD, E. P., CHILTON, A. B., GROTENHUIS, M., HÖNIG, A., JAEGER, T. A., and EISENLOHR, H. H., (Ed.), (1968). *Engineering Compendium on Radiation Shielding*, Springer, Berlin, 537 pp.

JUDEIKIS, H. S., (1968). 'Free Radical Yields in PTFE as a Basis for a Radiation Dosimeter', *Radiat. Res.*, **35,** 247—262.

KASTNER, J., (1969). 'Permanent Damage to TLD Meters by Fast Neutrons', *Hlth Phys.*, **17,** 368.

KEMP, L. A. W., (1969). 'Transistorised 'Doorpost' Radiation Alarm Monitor', *Acta radiol.*, **8,** 433.

KOEHLER, A. M., (1967). 'Dosimetry of Proton Beams with Small Silicon Diodes', *Radiat. Res. Suppl.*, **7,** 53–63.

LANDAUER, R. S., (1968). 'Development of Film Dosimetry', *Hlth Phys.*, **14,** 57–58.

LAUGHLIN, J. S., VACIRCA, S. J., and DUPLISSEY, J. F, (1968). 'Exposure of Embalmers and Physicians by Radioactive Cadavers', *Hlth Phys.*, **15,** 451–456.

LAW, J. J., (1969). 'Measurement of High Dose Rates with Aqueous Silver Nitrate', *Hlth Phys.*, **17,** 338–340.

MRC, (1960). Hazards to Man of Nuclear and Allied Radiations', CMND 1225, HMSO, London, 154 pp.

MINISTRY of LABOUR, (1964). 'Code of Practice for Protection of Persons Exposed to Ionising Radiation in Research and Teaching', HMSO, London, 64 pp.

NCRP, (1968). 'Medical X-ray and Gamma Protection for Energies up to 10 MeV—Equipment Design and Use', *NCRP Rep.*, No. 33, Washington.

NAGARAJAN, P. S., and KRISHNAN, D., (1969). 'Neutron Personal Monitoring', *Hlth Phys.*, **17,** 323–329.

OLLER, W. L., MENKER, D. F., and DAUER, M., (1969). 'Evaluation of the Fricke Dosimeter with Other Systems', *Hlth Phys.*, **17,** 653–659.

ORTON, C. G., (1966). 'Red Perspex Dosimetry', *Physics Med. Biol.*, **11,** 551–568.

PEIRSON, D. H., (1968). 'Neutron Dosimetry in Radiation Protection', *Physics Med. Biol.*, **13,** 69–78.

PENFIL, R. L., and BROWN, M. L., (1968). 'Genetically Significant Dose to the U.S. Population from Diagnostic Medical Roentgenology', *Radiology*, **90,** 209–216.

REES, D. J., (1967). *Health Physics*, Butterworths, London, 242 pp.
RYTILÄ, A., and SPRING, E., (1969). 'Radiation Dosimetry by Quinine Sulphate', *Hlth Phys.*, **17**, 336–338.
SCHULMAN, J. H., ATTIX, F. H., WEST, E. J., and GINTHER, J., (1960). 'A New Thermoluminescent Dosimeter', *Rev. scient. Instrum.*, **31**, 1263–1269.
SHAMBON, A., and CONDON, W., (1968). 'Lithium Fluoride Pellet Dosimeter', *Physics Med. Biol.*, **13**, 653–655.
SPURNY, Z., (1969). 'Bibliography of TLD', *Hlth Phys.*, **17**, 349–354.
TOCHILIN, E., GOLDSTEIN, N., and MILLER, W. G., (1969). 'Beryllium Oxide as a Thermoluminescent Dosimeter', *Hlth Phys.*, **16**, 1–7.
TOOLE, J. M., HENISCH, H. K., and MIYASHITA, J., (1967). 'Thermovoltaic Radiation Dosimetry', *Nature, Lond.*, **213**, 698–699.
VENNART, J., (1969). 'Radiotoxicology of Tritium and Carbon-14 Compounds', *Hlth Phys.*, **16**, 429–440.
WHITTAKER, B., and LOWE, C. A., (1967). 'Photosensitivity of Clear Perspex Dosimeters', *Int. J. appl. Radiat. Isotopes*, **18**, 89–91.
ZIEMER, P. L., (1968). 'Personal Radiation Dosimetry—Summary of Legal, Medical, and Administrative Needs and Problems', *Hlth Phys.*, **14**, 1–16.

Chapter 7

THE EFFECTS OF IONISING RADIATIONS ON LIVING ORGANISMS

7.1 INTRODUCTION

This chapter studies what has become known as radiobiology or radiation biology, the effects of radiation on living organisms. Unlike radiotracer methodology, which is relatively straightforward, radiobiology is a difficult subject because there are many steps between the initial impact of radiation and the final observed effects in the whole organism; precisely what happens at each step is not always fully known, and the correlation of molecular with cellular and whole-body changes is not easy. Radiobiology at the cellular level has difficulties enough and, at the higher levels of biological organisation, almost every statement has some qualification attached to it: the study of radiobiology can therefore very quickly become complex and confusing, and it is easy to get out of one's depth and become lost in the great mass of experimental data that has been accumulated. It is difficult to steer a middle course between the perils of over-simplification and over-complication.

The so-called *radiomimetic* substances, such as the nitrogen mustards, have been of some use in the study of radiation effects. They produce the same end-results as radiation, but the mechanisms by which the effects are produced may well be quite different for the two agents, and there is no way of telling at what point, if ever, the two pathways meet; there is indeed evidence that the primary reactions giving rise to chromosome damage are different for radiomimetics and for ionising radiation.

Radiation effects are seen over a wide time scale, from a fraction of a second to many years, and they are conveniently summarised in Figure 7.1. Immediate and early effects will be described first, from the molecular to the whole-organism level, followed by long-term and delayed effects.

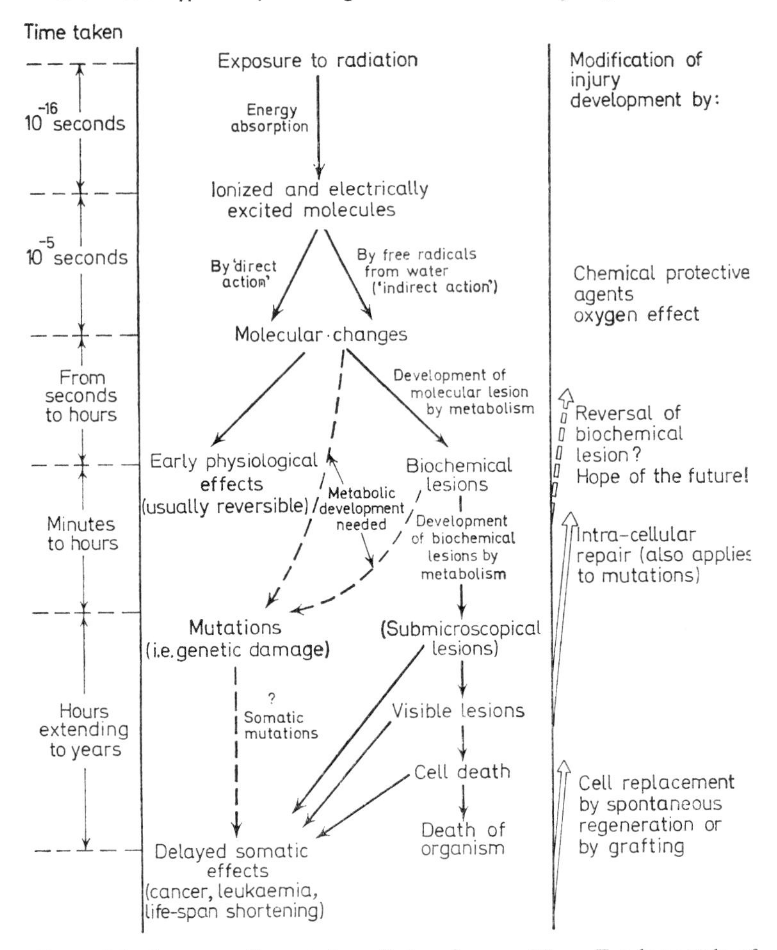

Figure 7.1. Sequence of events in radiation damage. (From Fundamentals of Radiobiology, *by Z. M. Bacq and P. Alexander, (1961). Courtesy Pergamon)*

7.2 MOLECULAR EFFECTS AND RADIATION CHEMISTRY

7.2.1 Direct and indirect action

Radiation may damage biologically important molecules in two ways. Direct action, as the name implies, is the alteration of a biological molecule through deposition of energy in it as the result of a primary interaction with the radiation. Indirect action takes place when the primary action of the radiation is on water, the biological molecule then being attacked by the highly reactive products of that radiolysis. A biological system suffers mostly indirect action since it is mostly water; however, action on water of hydration of a molecule is regarded as being direct rather than indirect.

The two actions may be distinguished by the *dilution effect:* the damage to a sensitive molecule in solution from direct action is proportional to its concentration, whereas that from indirect action is unaltered by dilution over a fairly wide range, as the concentration of water here stays relatively constant. While this test works perfectly well with solutions of, e.g., a virus or an enzyme, it is obviously not applicable to finding out whether radiation acts directly or indirectly within the cell for, though a cell suspension may be diluted, the cell interior cannot. Direct action appears to be more effective in a solution than indirect, but *in vitro* results cannot be extrapolated to conditions *in vivo*, since protection and repair mechanisms may exist within the cell that are not found in purely *in vitro* systems.

7.2.2 Ionisation, excitation, and free radicals

The first step in the chain of radiation damage is the formation of ionised and excited molecules by the transfer of radiation energy to them: this takes about 10^{-16} s. One ionisation event is generally assumed to require the transfer of about 33 eV of energy, but the actual amount needed to form an ion pair, the ionisation potential, may be much lower, 10–25 eV. The extra energy, about 8–23 eV, appears as excitation energy of the molecule. The current view is that chemical and biological radiation effects arise more from groups or clusters containing an average of three ionisations, rather than from single events, so that a 'primary ionisation event' is taken to be equivalent to the transfer of 100 eV of energy. Molecules may also become excited without ionisation, by absorbing a lesser amount of energy from the radiation. Excitation seems relatively unimportant in causing chemical and biological damage, although it has been found recently that as little as 5 eV of energy is able to cause significant changes in disulphide bonds.

Ionisation may be represented as: $A \rightarrow A^{+} + e^{-}$. The electron is soon captured, according to the relative electron affinities of the neighbouring molecules: $e^{-} + B \rightarrow B^{-}$. Oxygen is here important in biological systems, for it has a high electron affinity and is therefore a commonly found electron acceptor. The ion pair formed by the action of radiation is, of course, the positive ion A^{+} and the electron, though it is sometimes regarded (incorrectly) as being A^{+} and B^{-}.

The chemical effect of radiation, the *radiation chemical yield*, is expressed as the *G value*, the number of molecules changed for each primary ionisation event or 100 eV of energy transferred to the system. An alternative form of expression is the *ionic yield*, M/N, where M is the number of molecules changed and N is the number of ionisations. The two are related, G being equal to 3 M/N. G values (or ionic yields) have a very wide range, large values being explained by the occurrence of chain reactions.

The ions and excited molecules resulting from radiation action produce free radicals, which are highly reactive species (in the chemical, not the biological, sense) because of their distinguishing feature, an unpaired or lone electron in the outermost shell. They have a very short lifetime (about 10^{-5} s in water) and react within a very short distance, some 50 μm, of the site where they are formed. They should be carefully distinguished from ions, for confusion may easily arise between the unpaired electron of a free radical and the charge of an ion. Free radicals are written with a superscript dot or small circle, to represent the free electron; ions have the sign and magnitude of the charge they carry, thus:

$H^{\cdot}$	$OH^{\cdot}$	H^{+}	OH^{-}
Free Radicals		Ions	

7.2.3 Radiation Chemistry

The radiochemistry of *water* is important in the biological action of radiation, since water makes up such a large part of living things. The most likely route for the radiolytic breakdown of water is generally regarded to be that the primary ionisation:

$$H_2O \rightarrow H_2O^{+} + e^{-}$$

followed by the capture of the electron by a water molecule:

$$e^{-} + H_2O \rightarrow H_2O^{-}$$

each H_2O ion then breaking down into an ion and a free radical:

$$H_2O^{+} \rightarrow H^{+} + OH^{\cdot}$$
$$H_2O^{-} \rightarrow OH^{-} + H^{\cdot}$$

The H^{+} and OH^{-} ions then combine to give water, leaving the radicals $H^{\cdot}$ and $OH^{\cdot}$, so that the overall reaction is:

$$H_2O \rightarrow H^{\cdot} + OH^{\cdot}$$

A less likely alternative route is the primary ions recombining to give excited H_2O, this then breaking down to the same two free radicals $H^\cdot$ and $OH^\cdot$.

Many free-radical reactions are possible. The $H^\cdot$ and $OH^\cdot$ may recombine with each other, re-forming water, when no biological damage results, or giving H_2 (from $H^\cdot + H^\cdot$) or H_2O_2 (from $OH^\cdot +$ $OH^\cdot$). The last reaction is important because H_2O_2 is an active oxidising agent; it is also formed in the presence of oxygen by the reaction:

$$H^\cdot + O_2 \rightarrow HO_2^\cdot$$

forming the peroxyl radical, which can yield hydrogen peroxide by two reactions. Two radicals may combine directly:

$$2HO_2^\cdot \rightarrow H_2O_2 + O_2$$

or one radical may take up an electron from oxidisable material:

$$HO_2^\cdot + e^- \rightarrow HO_2^-$$

forming the anion of H_2O_2, which in all but strongly alkaline solutions picks up a hydrogen ion to form the peroxide. The $H^\cdot$ formed in the radiolysis of water is reducing, tending to lose its electron; the $OH^\cdot$ and $HO_2^\cdot$ tend to gain one, and along with the H_2O_2 are oxidising, so that overall the system is oxidising.

All of the above agents may attack and damage sensitive biological molecules, depending on how close they are to the site where the free radicals were created. The end-products may be formed in a single step, or by means of a series of reactions involving several intermediates. *Chain reactions* may occur, in which a radical generates another that can react in the same way, e.g. a radical may react with oxygen, forming a highly reactive and damaging peroxy-radical ($RO_2^\cdot$); this may remove hydrogen from another molecule and in doing so form another radical that is able to react with oxygen, and so on:

$$RO_2^\cdot + XH \rightarrow ROOH + X^\cdot; X^\cdot + O_2 \rightarrow XO_2; \ldots \text{etc.}$$

Much damage may be caused before the chain is terminated.

The occurrence of a chain reaction may be recognised by means of the yield, the extent to which solute molecules react with the free radicals from a given radiation dose. The yield increases with solute concentration up to about 10^{-3} to 10^{-2} mol/l, and then stays constant for all except chain reactions, in which the yield increases with solute concentration over a much wider range. Reactions of free radicals will continue until they have all been removed, either by combination with each other or by picking up a free electron, but, whatever reactions take place, the products are likely to affect the chemical and physical properties of the system, usually adversely.

The readiness with which a molecule is attacked by a free radical increases with its reactivity, its concentration, and its size. The important principle of *protection* of one molecule by another (Section 7.5.5) emerges from this: although a particular compound may be vulnerable to attack by reason of its reactivity, more radical attack may in fact be directed to another which is present in higher concentration and/or larger, so that the first is attacked less than it would be if the second were not present.

Stinson and Moore (1966) describe a histochemical technique for the direct observation of radiation reactions.

7.2.4 Radiation effects on biological molecules

A variety of changes may be observed if solutions of biologically important substances are irradiated, but extrapolation of these results to the living cell is hazardous because the situation is so very different: chemical and physical states are unlikely to be the same, and molecules that are radiosensitive *in vitro* may be protected *in vivo*.

Radicals may react with inorganic ions, e.g. Cl^-,:

$$OH\cdot + Cl^- \rightarrow OH^- + Cl\cdot$$

forming a chlorine radical, which will in turn attack other molecules.

The main reaction with organic compounds is removal of hydrogen, forming organic free radicals:

$$RH + OH\cdot \rightarrow R\cdot + H_2O$$

or

$$RH + HO_2\cdot \rightarrow R\cdot + H_2O_2$$

which may dimerise, polymerise, rearrange, oxidise, etc. The susceptibility of polymers to oxidation is increased by irradiation.

The formation of the organic radical may be reversed, i.e. repair may take place, by two reactions:

either

$$R\cdot + H\cdot \rightarrow RH$$

or

$$R\cdot + XH \rightarrow RH + X\cdot$$

Free-radical reaction is terminated in the first, but not in the second. Another possibility with organic compounds is the addition of free radicals across a double bond.

In general, it is found that: first, aromatic compounds are more resistant to radical attack than aliphatics, because they are resonant structures; secondly, a ring stabilises a chain; and thirdly, energy may react in a molecule elsewhere than at the site where it was absorbed, or be given off as heat, or it may be transferred to other molecules, though with less efficiency than its transfer within the same molecule.

Macromolecules may undergo structural alterations, which may be manifested as a change in physical properties affording a sensitive means of detecting a reaction affecting only one or two atoms in a molecule containing several thousand, though the chemical nature of the change cannot generally be determined at this level. The solution may change in viscosity, showing either an increase or a decrease; the zeta potential or surface charge carried by colloidal materials is very sensitive to radiation, and changes in it can be detected with radiation doses that give a biological effect but are without detectable chemical action.

Cross-linking—the formation of chemical bonds—is another effect of radiation. In dilute solutions, it is predominantly intramolecular, and may result in the molecule being pulled together to occupy less space, so that the viscosity of the solution is reduced. At higher concentrations of the solute, cross-linking is mostly intermolecular, between different molecules. As the radiation dose increases, the number of cross-links increases, and the solution gradually becomes more viscous, eventually turning into a gel; the gel is first detectable when there is an average of one cross-link per molecule, and the dose required for this is termed the gel point dose.

Breakage or degradation of a macromolecule may affect the primary structure, breaking it into pieces of various sizes. Such *main-chain scission* often occurs at one particular bond, suggesting that energy is transferred along the chain from the site where it was absorbed to the weakest link (*see* the third point above). Disruption of the secondary and higher orders of structure may also occur, e.g. through the breakage of disulphide bridges, as in proteins, or very often by the rupture of weak hydrogen bonds.

With *proteins*, direct action of radiation is considerably more efficient than indirect: as little as 50–200 eV of energy deposited directly in a protein can inactivate it, whereas some 50–200 hydroxyl free radicals are needed indirectly, although one radical may be enough if it hits the right place. Indirect action may also produce important changes, e.g. in solubility. All enzymes can be inactivated in solution; the doses required vary widely, and some protection is afforded by sulphur compounds. However, alteration of protein structure and function does not seem to be greatly important in primary radiation effects on biological systems.

In *nucleic acids*, several significant changes take place, both *in vitro* and *in vivo*, though in the cell there is some protective effect from the proteins associated with the nucleic acids. Mutations may arise through loss or alteration of bases: pyrimidines are more sensitive than purines, and the formation of dimers between adjacent pyrimidines, particularly thymine, seems to be important. Breakage of hydrogen bonds is likely to be significant only if several are broken at once in the same region; the inherent rigidity of the structure, unlike that of proteins, holds it together while hydrogen bonds are broken and allows them to re-form correctly. Breakage of the main strands may occur; a break in one chain only will be difficult to detect since (again) the DNA helix is inherently rigid and will hold together. The broken ends usually rejoin, but may not if oxygen is present so that they become peroxidised. Breaks may join up to form cross-links, either within the same DNA molecule, between two DNAs, or between DNA and protein. If there is a break in one chain within less than about five nucleotide units of a break in the other, a 'double break' arises, and the molecule separates into two pieces since there are now too few hydrogen bonds between the breaks to hold it together. Such a double break arises much less commonly than a breakage in one chain only: it is estimated that there is a 1 in 70 chance that two single breaks formed at random will come close enough together to make a double break. Alternatively, a single particle with high LET may cut through both chains at once, for which about 600 eV are required.

The chief reaction of *lipids* concerns the poly-unsaturated fatty acids, and affects the carbon atom between the double bonds: hydrogen is removed, and a resonating structure is formed. Peroxyl radicals and organic peroxides are readily produced in the presence of oxygen, often by means of free-radical chain reactions affecting from 20 to 1000 molecules.

Radiochemical studies on *carbohydrates* are few and inconclusive.

7.2.5 The primary radiation lesion

The primary radiation lesion is a change in a biologically important molecule that is induced by free-radical action very soon after, or even during, irradiation. So long as it is confined to this molecule and does not affect others, it is merely latent; it will not (and cannot) harm or damage the cell unless and until it is magnified, or multiplied, to the extent of interfering with the normal working of the cell. Metabolism is therefore essential for the expression and development of the primary lesion. How much metabolism must take place for the lesion to show itself depends on the molecule initially affected. Most kinds of molecule are present in the cell in large numbers and, although the compound may be very important, e.g. ATP, a change in a few molecules will go completely unnoticed, so that considerable metabolism is needed for the lesion to show itself. On the other hand, some molecules have few or no 'spares', as it were, notably the DNA storing the cell's information for reproduction and protein synthesis; there are such a small number that all, or nearly all, must be present for normal functions to be maintained. But in both cases, so long as the damaged molecules are not required and the cell does not have to try to use them, it can function normally. What is important is metabolism of the lesion (the damaged molecule(s)), not the normal metabolism of the other unchanged parts of the cell; however, there is much direct or indirect molecular interaction in the cell, so that a change in one or two molecules may ultimately affect many more.

In contrast to the metabolism of the lesion is its repair, through which *recovery* from the radiation damage takes place, for by no means all lesions are permanent and irreparable, e.g. damaged molecules may be broken down and excreted, and there is evidence that a damaged section in a chain of DNA can be replaced.

Free radicals can undergo a great variety of reactions with biologically important molecules, and therefore the problem in answering the question of what is the initial biochemical lesion is not so much to find a biologically important free-radical reaction as to decide which of the many possibilities is the key reaction that starts off the chain of events leading to visible cellular damage. In answering the question, it is important to bear in mind that the only valid observations are those made within a short time of irradiation, before the metabolism of the lesion has had time to spread to many parts of the cell; otherwise it will be difficult or impossible to work out the original nature and site of the initial lesion. It is no use whatever noting the biochemical changes in cells that have developed a visible lesion, since the initial chemical lesion inevitably and invariably precedes it. In addition, there is little change in the general biochemistry of irradiated organisms, except during the last days of life, and the disturbances observed are secondary to the effects of radiation sickness.

A clue to the nature of the initial radiation lesion is given by the fact that a dose of 100 rad kills many cells, but inactivates only a few molecules in those cells, most being competely unaltered. It is also known that the RBE of radiation *in vivo* increases with increasing LET, though the RBE is not the same for different effects. The properties of the lesion must agree with these observed facts, and a number of possible alternative hypotheses have been proposed:

Radiation may affect *an essential molecule*, of which almost every one is vital to the normal functioning of the cell; an obvious choice is nuclear DNA. Key metabolic enzymes are less likely, since there are relatively many molecules of them, and they can be shown to be functional after a damaging radiation dose. Interference with DNA is almost certainly the most important cause of long-term and genetic radiation effects.

Radiation effects might follow the *interruption of energy supplies*: blocking of key enzymes might stop normal metabolism, so that the body would have to use other energy sources which might be quickly exhausted.

Radiation might *poison the cell*, either by direct alteration of particular compounds, or indirectly via the interruption of energy supplies above, by the accumulation of intermediary metabolites that are toxic in high concentration. Kuzin *et al.* (1966) suggest that the oxidation of orthophenols to orthoquinones is fundamental in the biological action of high-energy radiation. Glyoxal may be a cytotoxic agent resulting from the irradiation of carbohydrates.

Radiation might break down *intracellular barriers*. This, the so-called *membrane-damage or enzyme-release hypothesis*, postulates that the primary lesion caused by radiation is damage to the membranes around and within the cell, which can resist isolated ionisations but not the impact of several simultaneously. The consequent alterations in the structure of the membranes lead to changes in their permeability or (if sufficiently extensive) to their rupture: thus all kinds of molecules which are normally held on one side of the membrane can escape, including enzymes, which are then able to react in an uncontrolled manner with substrates they might not normally encounter. This theory explains better than others many of the observed immediate effects of radiation, and in this respect is now favoured as being the most likely to be true. Through it, the former 'nucleus versus cytoplasm' controversy over the location of the primary site of radiation damage has in part lost its meaning for, once intracellular barriers are broken, the released enzymes are free to travel, and the site of the primary lesion can no longer be identified with the site at which the enzyme causes damage. Cytoplasmic enzymes can act on the nucleus, as the nuclear membrane allows the passage of large molecules.

Radiation might induce the formation of a virus: Fischinger and O'Connor (1969) present evidence that x-irradiation can induce a leukaemogenic virus in mice.

But no one hypothesis can satisfactorily account for all the observed effects of radiation on the cell, and it is likely that all the mechanisms of damage in the several theories have some part to play at some point in space and time.

7.3 CELLULAR EFFECTS

7.3.1 Outline

The initial radiation lesion is essentially chemical in nature. Its manifestation as a cellular effect will depend on the types and numbers of molecules affected, and to some extent on their location in the cell, so that the several types of cellular effects from radiation are ultimately traceable back to defects in particular molecules; however, the correlation of molecular with cellular changes is not easy, because the cell is a highly organised and finely controlled system, and is much more than a random mixture or simple aqueous system of its component biochemicals.

Cellular effects are divisible broadly into three categories of interference: with the cell's vegetative function, with its information store, and with its division. Most, if not all, may arise from two main primary lesions, affecting *membranes* and *nucleic acids* respectively. Radiation damage to membranes may be reflected in a physiological effect or a biochemical lesion, or be expressed further as a *visible anatomical defect*, which if extensive enough may eventually result in the cell dying through breakdown of its vegetative metabolism. Damage to a nucleic acid may also appear as a reversible short-term (physiological) effect, through modification of its colloidal properties. But the real importance and significance of damage to nucleic acids, and hence to chromosomes, is that a part of the cell's store of information becomes garbled or lost, and that these effects are *long term* and may be *delayed action*, for they will be passed on to future generations if the cell divides; they may be regarded as *genetic lesions* or *mutations*, and are the most serious effects of ionising radiation.

7.3.2 Physiological effects

Physiological effects are seen early after irradiation and (with small doses at least) are usually reversible. They include some macro-scale phenomena that are undoubtedly due to permeability changes, such as the release of biogenic amines and interference with nerve functions; the nervous system undergoes several rapid physiological reactions to radiation, though anatomically it is unaffected and in this respect has been considered radioresistant. Light-emitting bacteria respond to radiation with a steady decrease in light output as long as exposure to radiation continues; when irradiation is stopped, light emission increases again, but levels off below the pre-irradiation intensity, indicating that a residue of the damage caused by the radiation is not immediately repairable (Section 7.5.1). The uptake of certain ions by plant roots is in the same category: it is depressed by irradiation and thereafter soon returns almost to normal. A physiological effect involving DNA, chromosome 'stickiness', arises from an alteration of the surface properties of the chromosomes whereby they seem to adhere to each other if they happen to touch. They appear abnormal at metaphase and cannot separate cleanly at anaphase, and cell division is hindered or incomplete. The effect is reversible, since it does not appear if there is a delay between irradiation and division, the damage presumably having been repaired in the interim (Section 7.3.5).

7.3.3 Biochemical changes

Radiation affects enzyme activity, but so many different factors are involved that the picture is highly complicated. All enzymes can be inactivated in solution by radiation, and the doses required vary greatly, although they are usually much higher than those required to stop cell proliferation; the activity of some can be enhanced. In pure solution, enzymes are more radiosensitive than *in vivo*, whereas when attached to isolated mitochondria they are more resistant than *in vivo*. In addition, the interpretation of experimental studies is very difficult because there are many steps in the procedures, whereby many artefacts may be introduced. It is almost impossible to make a general statement without going into great detail, except that radiation effects on enzymes seem to be secondary following on from primary damage to something else. Enzyme studies have indeed provided some support for the membrane-damage hypothesis of the primary action of radiation, e.g. the relative concentration of catalase in the cell sap, as compared to that in the mitochondria, was found to be increased by irradiation, and administration of soya-bean trypsin inhibitor, which inhibits proteases, reduced early radiation lethality. Energy metabolism in certain types of cell is altered by radiation, phosphorylation, or the production of ATP being reduced, or uncoupled from oxidation. This too may be a secondary result of damage to the mitochondrial membrane. The synthesis of DNA may be slowed, probably as a result of delay in cell division (*see* below). Damage to the DNA template may lead to inability to synthesise a particular enzyme, and thence to interference with the metabolic pathway in which that enzyme takes part. This is of particular importance with gametes, whose one cell develops into a whole organism. Some biochemical changes, e.g. the level of β-amino-iso-butyric acid in the plasma, may be able to act as biochemical indicators of radiation damage, but overall the biochemistry of irradiated organisms is little altered, except near death, when the change is secondary.

7.3.4 Anatomical changes

Physical damage to membranes can be observed in the electron microscope following fairly large doses of radiation, which again supports the membrane-damage theory. The plasma membrane and the cytoplasmic structures appear more resistant than the nuclear membrane.

Structural aberrations appear in chromosomes after irradiation. Deletion of a fragment from the rest of the chromosome leaves a piece with no centromere that cannot in consequence take part in cell division; it forms a bridge at anaphase and results in the formation of a non-functional micronucleus. This loss of material may lead to death of the cell after one or two divisions. The fragment may attach to another chromosome, and persist through many generations; an exchange of fragments, rather akin to crossing over in meiotic division, may also occur. Various theories of breakage, inversion, rejoining, deletion, etc., of chromosome fragments have been developed, but it must be appreciated that breakage is certainly not a primary effect, for radiation cannot break a chromosome directly like a scythe cutting grass; a break is an anatomical lesion resulting from metabolism of a preceding biochemical lesion, although little is known of the chemical changes in between. Likewise, the rejoining of broken fragments is a complex metabolic process.

Chromosome damage has been observed with radiation doses lower than those needed for alteration in the leucocyte count of the blood (Section 7.4.3), and periodic study of the chromosome picture is therefore suggested to be important as an early check on radiation exposure.

There appear to be two schools of thought on how the sensitivity of chromosomes to structural alteration varies with the stage of division. According to Casarett (1968; p. 112), sensitivity is least in the G_1 stage of prophase (the resting, passive, non-synthesising stage), slightly greater in the G_2 stage, higher again in the S stage, between G_1 and G_2, when DNA is being synthesised in preparation for division, and greatest during metaphase and anaphase when the chromosomes are moving apart. At this stage, a mitotic cell is about three times, and a meiotic cell about ten times, as sensitive as a resting one. Bacq and Alexander (1961; p. 254) assert that the cell is most sensitive to chromosome damage in the resting stages, and is relatively insensitive at meta- and anaphase, although damage caused when the cell is resting will not in fact be observable until the next division, when the chromosomes are visible as individual structures.

7.3.5 Changes in cell division

The effects of radiation on cell division are best studied *in vitro*, since they can here be observed far more easily than in the whole organism. Colonies of unicellular organisms, or of cells taken from a multicellular organism and grown in tissue culture, are used. Recent developments in mammalian cell culture have provided a powerful attack on the problems of cellular and tissue radiobiology, making it possible to study cell population kinetics (Lamerton, 1968; Elkind and Whitmore, 1968). The investigation of the different phases of the cell cycle and of the dynamic characteristics of cell populations provides parameters through which the response of tissue to injury by radiation can be analysed. The labelling of cell cultures with tritiated thymidine gives an estimate of the turnover time, the time required to produce the same number of cells as those initially present, though this time may be the result of either the rapid division of a few cells or the slow division of many cells, and there is no way of telling which.

It is difficult, if not impossible, to extrapolate from tissue culture to the whole organism, because the conditions are so very different. The cells tend to become less specialised in form as the culture grows older, i.e. they de-differentiate; and cell types which normally never divide *in vivo* (except perhaps following injury) will do so *in vitro*. Irradiated cells in the intact organism may affect normal cells in another region, and vice-versa, and it is often very difficult to decide whether the observations are due to such *abscopal* effects or to direct effects of radiation. And, though a lesion may affect cell division *in vitro*, it is of no consequence *in vivo* if the cells never divide, for it will not then be metabolised (Section 7.2.5). A long time interval between cell divisions in the intact organism may enable the lesion to be repaired before division occurs (Section

7.3.7); on the other hand, the damage may be permanent, and may show itself when the cell eventually divides, much later though this may be. However, in spite of these difficulties, tissue culture of isolated cells is at present the most promising method of studying the effects that radiation may have on cell division in a multi-cellular organism.

The cells in a clone are descended from one parent and should therefore be all alike, but their precise response to a given radiation dose varies. If their division is random, the variations arise since the radio sensitivity of cells depends to some extent on the stage of division or of the cell cycle at the time of irradiation. If division is synchronised and all cells are at the same stage of division together, then chance decides which cells are damaged and where, by analogy with the random effect of firing a shot-gun into a flock of identical birds.

A problem of definition arises in the consideration of what is meant by saying that radiation kills a cell. A cell dies when it ceases to metabolise and its ordered structure is broken down, but is it to die at once, after a period, or after a few divisions? In cancer radiotherapy it may be sufficient if the tumour cells are prevented from dividing, so that though the growth may still be alive, in the sense that the cells respire and use nutrients, it is effectively dead in that the cells cannot proliferate and the tumour cannot increase in size.

As with other effects, the severity of the radiation effect on cell division increases with the dose, though different cell types show great differences in radiosensitivity. The changes induced are four-fold: first, delay in division; secondly, death following one or a few divisions; thirdly, complete inhibition of division, with death following after a period of time; and fourthly, instant death. Much greater doses are needed for the third and fourth changes than for the first two.

Mitotic delay

Small doses of radiation produce a temporary alteration in the pattern of cell division, mitotic delay; as the name suggests, mitosis does not take place at the expected moment, but is delayed temporarily, probably by means of a reversible physiological effect (Section 7.3.2), a disturbance in the mechanism of separation of the chromosomes. Cells in particular phases of mitosis are held there for longer than normal, or may go back a stage or two. The cycle is most sensitive to radiation when division is about to start, in the G_2 stage of interphase; as little as 1 rad is all that is required to affect it in some tissue cultures (grasshopper embryo neuroblasts). Cells in prophase are also sensitive, up to a critical point in very late prophase: they either stop or revert to G_2. Irradiation after this point does not delay mitosis, and division proceeds normally, though chromosome stickiness may make the following stages a little slower than normal. Chromosome stickiness is not seen in delayed cells, because there has presumably been time for the lesion causing it to be repaired. Interphase cells in the G_1 and S stages are less sensitive. A low dose of radiation may leave them unaffected, but retard the G_2 and prophase cells, so that there is an accumulation of cells at the G_2 stage. As the delayed cells recover, both they and the unaffected cells, whose division is continuing as usual, all divide at once, so that, being brought into step, as it were, there is an apparent increase in cell division. This led to the old theory that radiation enhanced mitosis, which was disproved once the truth was realised, that radiation had in fact altered the pattern of timing of cell division. In some circumstances cells have been shown to accumulate at the end of G_1.

Mitotic delay may be expressed in terms of the *mitotic index* (*MI*), the proportion of cells in the culture which are dividing at any given time. Under suitably controlled conditions, the *MI* can be held constant, and mitotic delay then appears as a decrease in *MI*, followed by an increase as the compensatory wave of increased division takes place, which gradually falls off back to normal. The *MI* thus follows an S-shaped curve (Figure 7.2).

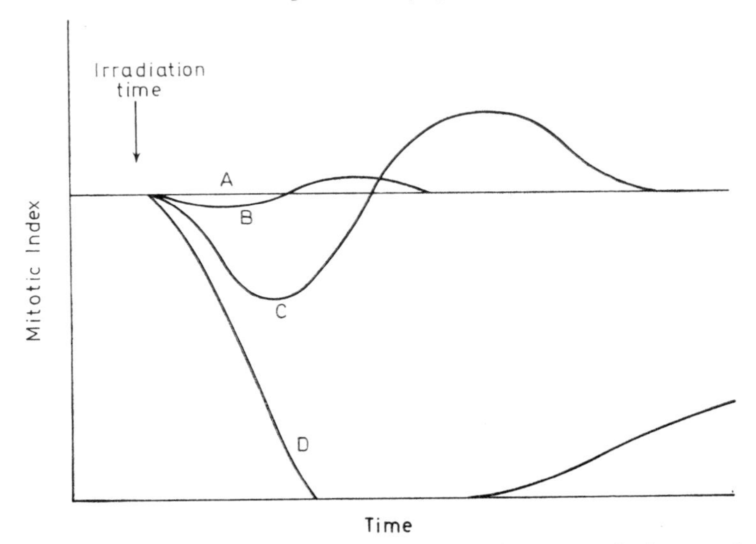

Figure 7.2. Mitotic delay of cells in tissue culture (diagrammatic): A, *normal, no irradiation;* B, *low dose;* C, *moderate dose;* D, *high dose*

The rate at which the dose in given is important in mitotic delay. Low rates are without effect, indicating that the cell can repair itself as fast as it is damaged. If a moderate dose is given to a culture of constant *MI* at a very high rate, the period of irradiation will be very short; not all cells will be caught at their most sensitive stage, and the effect will be less than the same dose given at a lower rate. In addition, unlike many radiobiological effects, there appears to be a threshold dose below which mitotic delay is not produced.

As mentioned above, mitotic delay is a temporary reversible effect, and the implication is therefore that it results from interference with a metabolic process. Inhibition of DNA synthesis has been suggested as a possibility, but it may be the result rather than the cause of altered mitosis; also it is not seen in all types of cells, and the most sensitive part of the mitotic cycle for it is not the same as that for mitotic delay. An alternative is damage to the mechanism controlling cell division or maintenance of growth.

Mitotic inhibition

The effect on mitosis becomes more pronounced with higher doses of radiation, which may not only delay division of cells but also prevent it altogether. The mitotic index returns to normal either very slowly or not at all (curve D in Figure 7.2) and never exceeds the pre-irradiation level, showing that mitosis has been inhibited rather than delayed, and that a proportion of cells in the culture have been irreversibly damaged, being now incapable of division or able to divide only more slowly. The slow increase in *MI* now results not from recovery of damaged cells, but rather from repopulation of the culture by multiplication of those surviving cells that have retained their ability to divide indefinitely. The fraction surviving can be estimated from a semi-logarithmic plot of cell concentration against time (Figure 7.3). The normal picture is a straight line (curve A); irradiation may produce, depending on the dose, an inflection, a horizontal portion, or a decrease in cell count, either temporary or permanent (curves B, C, D, and E respectively). Except in the last case, the culture eventually recovers and its multiplication rate returns to normal. The graphs may then be extrapolated backwards to zero time, and the ratio of the cell number at the intercept (B', C', D') to the original number (N) is a measure of the surviving fraction. It decreases as dose increases. Alternatively, the number of cells retaining the ability to form clones may be determined directly by plating them out and counting how many form visible colonies.

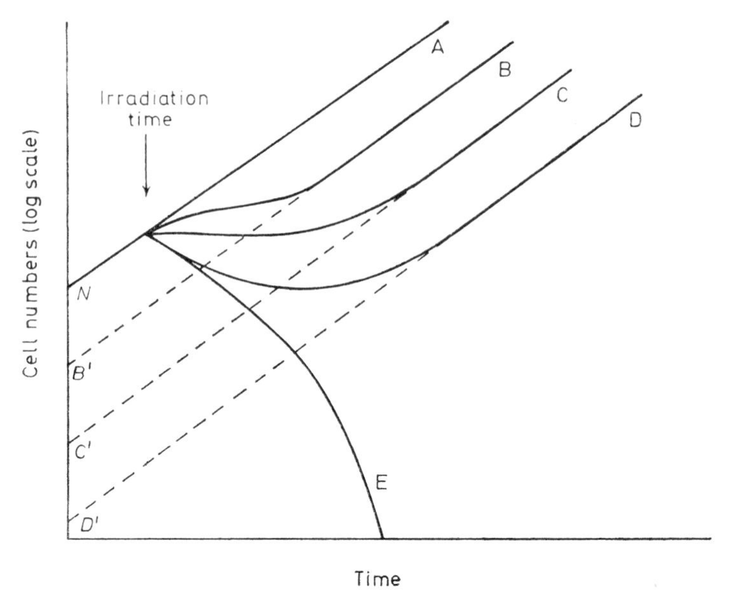

Figure 7.3. Effects of irradiation on cells in culture. For explanation of lettering, see text

Permanent impairment of the ability of a cell to reproduce indefinitely comes about through irreparable damage to the chromosomes, by breakage or otherwise. Though gross chromosome damage may or may not be apparent as a visible lesion, it need not necessarily inhibit cell division; some highly malignant tumours are anatomically very abnormal, and presumably the mechanism controlling their division has gone awry, with all restraining influences removed. There appears to be no direct relationship between damage which is manifested as mitotic delay and mitotic cell death, and there is no reason to suppose that delayed cells will be more likely to die than others.

Mitotic and interphase death

Mitotic death may follow deletion of fragments of the chromosomes. The cells divide once or a few times and no more, and may then develop into 'giant' cells up to about 30 times the normal volume (about three times the diameter) before they ultimately die. Cells suffering interphase death can never divide again after irradiation; they too may develop into giants in tissue culture, though they may not increase in size *in vivo*. In general, interphase death needs higher doses than mitotic death; cells vary widely in sensitivity, those with the lowest rate of division being most resistant, and some, e.g. lymphocytes and oocytes, being particularly sensitive. Cells of several mammalian lines in culture are most sensitive to prevention of proliferation at the end of the G_1 stage of the cell cycle.

Instant death

Very high doses are required to bring about instant or immediate death, i.e. breakdown of cells within a short time of irradiation. The primary lesion is almost certainly membrane damage; one probable mechanism is rupture of the membranes bounding the lysosomes, the small mitochondria that are the so-called suicide bags of the cell. They contain lytic enzymes which under normal conditions come into action only after the death of the cell, being then responsible for its autolysis, whereas following irradiation they may be released prematurely and begin to act while the cell is still alive, so killing it instead of acting only post-mortem. Experiments *in vitro* have shown that lysosomes are resistant to radiation, and that high doses, in excess of 10 000 rad, are needed to break them open.

7.3.6 The target theory

This theory was developed to provide a method by which the size and shape of the critical structure involved in the initial lesion could be calculated. Ionisation releases energy in a localised area where the bulk of the damage is concentrated, and the theory was based on the consideration that biological systems have a sensitive volume within which release of sufficient energy will bring about loss of function, and that the site of the biological end-effect or anatomical lesion is exactly the same as the site of this initial primary irradiation event. If one ionisation was needed for inactivation, a one-hit process was involved; a multi-hit process required more than one target to be hit, i.e. there were a number of smaller sites distributed in the cell rather than only one larger target. The sensitive volume, i.e. target size, and the number of hits required for inactivation were calculated from dose–response curves.

The basic assumptions are simple, and probably in part correct, but many conditions must be fulfilled for the theory to be valid and, even so, much guesswork is involved. In any case, it applies only to direct action, and it is relevant here in the study of the size and shape of biologically active molecules and agents or sites thereon, such as DNA and viruses. It cannot be used for indirect action via free radicals, so that it is inapplicable to most biological phenomena, for which, in fact, its predictions are completely irreconcilable with the observed facts. The membrane-damage hypothesis (Section 7.2.5) means that the initial lesion and its observed effect need not be in the same place, and metabolism and repair of the lesion pose an insurmountable barrier to calculation of target sizes. The target theory approach is therefore without value in helping to understand the nature of the primary sites of damage, and it is meaningless for cellular radiobiology.

7.3.7 Comparative radiosensitivity of cells

Almost from the beginning of radiotherapy, it was recognised that different types of cell do not all have the same sensitivity to radiation. Recognising this, *Bergonié and Tribondeau* put forward in 1906 the Law that bears their name; it states that the sensitivity of cells to irradiation is directly proportional to their reproductive activity and inversely proportional to their degree of differentiation. Thus, unspecialised cells which have not differentiated, e.g. those in lymph nodes, spleen, thymus, and bone marrow, are most sensitive. Dividing tissues are less so, e.g. testes, ovaries, mucous membrane, and skin; the dose required to kill 50% of a population (LD_{50}) of skin cells is 5000 rad, compared with 300 rad for lymphocytes. Non-dividing or 'static' tissues are most resistant, such as cartilage, bone, lungs, nerve, muscle, and the viscera. However, comparison of the sensitivity of different cell types is difficult because degree of specialisation cannot be defined absolutely, but only relatively within one line of differentiating cells.

An important question is the criterion used to measure sensitivity. It is an inadequately defined concept, and has been said to be capable of meaning all things to all men; in addition, it is affected by the experimental technique and the nature of the radiation, and it has led to great confusion (Lamerton, 1968). If interference with some aspect of cell division is taken as the parameter of sensitivity, then non-dividing tissues may appear resistant simply because they do not divide and the radiation lesion is unable to manifest itself. However, if slowly dividing tissues are more resistant than those dividing quickly, it seems probable that they are so because the lesion has time to be repaired in the interim between irradiation and division, and that, if and when a normally non-dividing tissue should eventually divide at some later date, it would be expected (though it is not certain) that here too the radiation damage would have been repaired.

Bergonié and Tribondeau's Law is used as the basis for the radiotherapy of rapidly dividing cancers, but it does not apply to those that do not divide appreciably faster than the surrounding healthy tissue. The oxygen effect (Section 7.5.2) and other factors may be employed to increase the sensitivity of cancers to radiation.

Cell sensitivity also differs within a population of cells, such as a bacterial culture. If the fraction of a population surviving is plotted logarithmically against the dose, two types of straight-line response appear, one with and one without a shoulder (Figure 7.4).

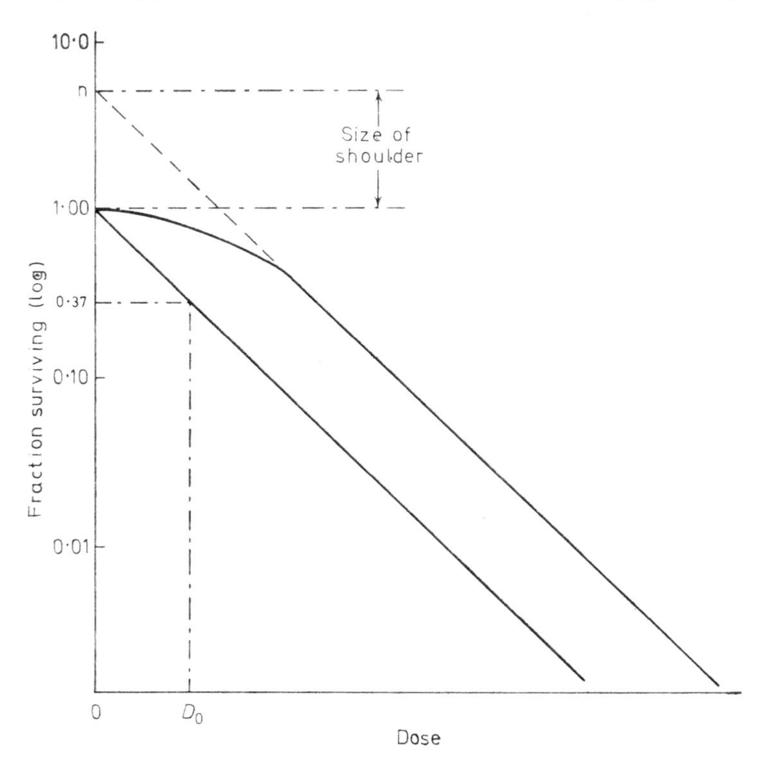

Figure 7.4. Bacterial dose–response curves

They show that, although a few bacteria are killed by low doses of radiation, a few survive high doses, and it is these that are the most important in any consideration of sterilisation (Section 5.1). Radiosensitivity of bacteria is therefore defined in terms of the dose–response line, by specifying the slope and the size of the shoulder, and not as a 50% lethal dose (LD_{50}); in addition, different curves could give the same LD_{50} value. Lethal doses are, however, very useful for multicellular organisms, whose dose–response relationship is quite different.

The size of the shoulder, or *extrapolation number*, n, is found by extrapolating the straight line part of the graph to zero dose, and the slope is expressed as the D_0 or D_{37} value, the dose required to reduce the viable cell number to 37%, i.e. to kill 63%. These numerical factors are derived from target theory, whose terminology has remained. Mammalian cells in culture are very much more sensitive than bacteria; their D_0 values are usually between 100 and 200 rad for most cell types, with a few just below or above these limits. Although cell culture conditions are very artificial, the agreement between the figures obtained for mammalian cell types tested both *in vitro* and *in vivo* has been remarkably good.

7.4 WHOLE-BODY EFFECTS OF RADIATION

7.4.1 Outline

The whole-body effects of radiation vary not only with the total radiation dose but also with the rate at which it is delivered, and the study of the dose–response relationship is complicated further by the variation being in the nature of the response as well as in its magnitude.

Whole-body effects are the end-results of a chain of events, but their interpretation in terms of molecular and cellular effects is very difficult, for in the same way that the cell is more complex than a mixture of its component biochemicals, so the organ and the body are far more than the mere sum of the cells that make them up. The difficulties of interpreting radiobiological effects in vertebrates increase the value of observations on material which is not subject to neurohumoral regulation.

Mice are among the most commonly used subjects for the overall study of the whole-body effects of radiation. The picture obtained from the results of work on these and other small laboratory animals is likely to be true for most, if not all, other mammals also, but different animals do vary in their response to radiation and, though the same general principles may apply, it is well to be very wary of extrapolation from one animal to another, or from animal results to humans, in too great detail.

The effects of radiation on the whole body may be seen in the short term, immediately after or within a few days of irradiation, as the acute radiation syndrome. In addition, long after recovery from these early effects is complete and they have been forgotten, the individual may suffer long-term or delayed effects; there may also be genetic effects that are passed on to future generations without affecting the individual. These late effects arise also from continuous low-intensity chronic irradiation or a succession of small sub-acute doses.

Low doses may produce only delayed effects, with none in the short term; there appears to be a threshold of about 25–50 rad in Man, and rather more in animals, below which no early symptoms are seen. It was once believed, from extrapolation of high-dose experiments, that radiation had no effect in the long term below a certain minimal threshold dose, but this extrapolation is dubious since the experimental uncertainty increases as the dose gets less, and it is not known whether the dose–response graph is a straight line in this region. The causes of short- and long-term effects may well be different, and information about one cannot be taken to have any bearing on the other. Short-term effects depend largely on dose rate, indicating that they are, if not too severe, repairable, whereas late effects do not, so that they are irreparable and therefore permanent and cumulative. It is now accepted that there is *no known threshold* for long-term radiation effects; any dose of radiation, however small, is capable of causing some damage, though it is not certain to do so, and the chances of escaping unscathed increase greatly as the dose gets smaller.

The data on late radiation effects are incomplete and not clear-cut, and it is difficult to draw definite conclusions, because the interpretation of experimental results is complicated by several factors as follows:

1. Accurate measurement of small doses is difficult;
2. The dose rate has some significance, for some repair is possible in immature germ cells, though, once a gene has mutated, it is permanently altered;
3. It is difficult to ensure that the effect obtained is actually due to the radiation and is not caused by anything else;
4. The increases in the incidence of injuries are very slight, requiring strict statistical treatment, and could necessitate the study of a very large population to be sure that the increase is real and not a chance fluctuation;
5. It is always difficult to extrapolate from animal experiments to Man;
6. The time scale.

But, although there may be these uncertainties, the fact remains that long-term effects are cumulative, need only low doses, have no threshold, and may be genetic; and this is why they, rather than the acute radiation syndrome, are considered to be the *most serious effects* of radiation, which decide the maximum permissible levels of radiation exposure (Sections 6.5.1, 6.5.4). Of these effects, malignant disease is the most important.

The effects of different doses and dose rates of radiation may be broadly summarised as in Table 7.1; the figures are approximate, to indicate relative orders of magnitude, and are not exact.

Table 7.1. EFFECTS OF IRRADIATION AT DIFFERENT DOSES AND DOSE RATES

ACUTE IRRADIATION

Dose (rad)	*Effect*	
100 000	Spastic seizures; death in minutes	Widespread disorganisation and death
10 000	Damage to central nervous system (CNS); death in hours	
1 000	Circulatory changes; death in days	
100	Radiation sickness; decreased life expectancy and disease resistance; sterility	
10	Cataract; impaired foetal development; other symptoms not obvious, or undetectable (calculated risk?)	

CHRONIC IRRADIATION

Dose (rad/day)	*Effect*	
10	Debility in 3–6 weeks; death in 3–6 months	Extrapolated from animal data
1	Debility in 3–6 months; death in 3–6 years	
0·1	Reduced life expectancy; symptoms appear after several years; acceptable dose 1930–1950s	
0·01	Currently acceptable dose	
0·001	Natural radiation	

7.4.2 Death

There is no doubt that a fairly large quantity of radiation, delivered quickly as a single dose, will kill an animal, in the sense that it would have been most unlikely to have died from natural causes otherwise, so that the statistical probability is overwhelming that its death is directly attributable to the radiation. This dose is about 700 rad (minimum) in mice. At lower doses, the probability of death rapidly decreases, and a variety of short-term effects appear; if the animal survives or escapes these, as happens with low doses or at very low dose rates, a long-term effect may appear in the shortening of its life expectancy compared with a normal animal, although it is now a problem of semantics to say whether in fact radiation has killed the animal.

Death may result from damage to any of the various organ systems of the body; they vary in radiosensitivity, and the number damaged by a given radiation dose increases with the dose. Death following a high dose is not necessarily the result of damage to the tissues of greatest radiosensitivity; although such damage will undoubtedly occur and would in time be fatal, damage to other tissues may have a more immediate effect and be in fact responsible for killing the animal. The time between irradiation and death varies with the system injured and the severity of the injury, and it may be predicted that, of the systems damaged by a given dose, the system for which this time is the shortest will be responsible for death of the animal, and the symptoms seen at death will therefore relate to this system. Accordingly, the cause of death, and the symptoms seen, will be different in different dose ranges. This is borne out fairly well in practice, although there is the complication already mentioned that all the organs in the body are part of an integrated whole and interact with each other and, if the animal survives more than a few days, it becomes increasingly difficult to decide what is the exact cause of death.

Massive irradiation of mice (tens of thousands of rad) produces near-instantaneous death, the result of the simultaneous disruption of the whole of the body's chemical activity. It is known as the *hyper-acute syndrome*, *molecular death*, or *immediate death*. With doses of about 15 000 rad, the mice survive for 1–2 h, and the symptoms indicate that derangement of lung function, the *lung syndrome*, is the cause of death. A somewhat lower dose, 10 000 rad, gives a different picture, the *central nervous system syndrome*, in which hyper-activity and disturbance of the CNS kills the animal after 1–2 d. Post-mortem examination of these two syndromes shows haematological and gastro-intestinal damage also, which would itself have proved fatal had not the animal died before these lesions had had time to develop. In other words, although a relatively radiosensitive organ system has been damaged sufficiently to kill the animal, interference with a relatively radioresistant one is the actual cause of death.

Over the large dose range 900–10 000 rad, the *gastro-intestinal syndrome* kills the animals within 3–5 d. The cells lining the gut are killed, and food materials cannot be absorbed, although the cause of death is not so much malnutrition as the severe loss of fluid and sodium through the damaged gut.

Doses of 300–900 rad (200–700 rad in Man) result in an overall picture of radiation injury to several systems, and death is not a certainty. The animals which do succumb may die (after about 10–15 d) of infection and blood loss resulting from the *haematopoietic* syndrome, or from a secondary cause that becomes increasingly difficult to determine the longer the animal survives. Lower doses give the same picture, with less severe symptoms, but there are virtually no deaths, so that over quite a narrow range the relationship of radiation dose to survival time alters profoundly, the latter increasing sharply from a few percent to something approaching 100% of normal; Figure 7.5, the response of experimental animals to irradiation in terms of survival time, shows this very clearly.

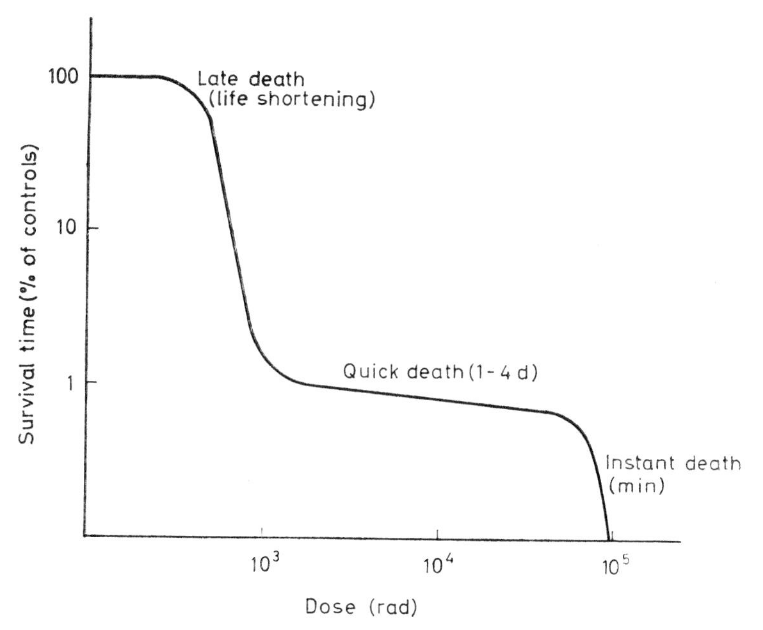

Figure 7.5. Whole-animal survival time after irradiation

Herein lies the justification for one widely used method of expressing the response of animals to radiation; this is the *number surviving for 30 days* after irradiation, or alternatively the LD_{50} (30 d), i.e. the dose needed to kill half a group of animals within 30 d. At first sight, this seems a very arbitrary criterion, for a 29–d survivor would be scored as a death (a 'hit'), whereas one dying after 31 d—lasting only 7% longer—would count as a survival (a 'miss'). But the sudden alteration in the relation of dose to survival time means that a line can be drawn between immediate or early death and long-term survival. Very few animals die after 3 weeks, and the figure of 30 d is chosen as a time which is well on to a plateau region of slow variation, so that it does indeed have sound foundation. An animal lasting a month after irradiation is likely to live out something approaching its normal life-span, and radiation which does not kill within 30 d is thus regarded as being not acutely lethal.

7.4.3 Radiation Sickness

More important than death in this low dose range of a few hundred rad is the syndrome of acute radiation sickness, whose symptoms, and their severity, depend to a large extent on both the total dose and also on the rate at which it is delivered. Because radiation sickness is of direct importance to Man, the literature abounds with descriptions of symptoms and observations, and their interpretation, quantification, and alleviation; and only a brief general outline of the syndrome can be given here. Experimental work is confined to animals for obvious reasons, and once again the dangers inherent in extrapolation to Man should be stressed, though the documentation on human exposure through accidents shows that the syndrome is very similar in both animals and Man (e.g. Lushbaugh *et al.*, 1967, 1968). The CNS is most involved with doses of more than 2000 rad; between 500 and 2000 rad the effects on the gastro-intestinal system are of major importance, and below 500 rad those on the haematopoietic system. The LD_{100} in Man, at which death is a certainty, is about 800 rad; the LD_{50} is about 300 rad; and the LD_0, the dose just insufficient to kill any subjects, is about 200 rad. But one great difference from animals, and perhaps the most dreadful thing about radiation sickness from the human point of view, is the mental anguish of a victim who understands radiobiological mechanisms and who therefore knows what may happen to him; here indeed is science horror fiction made real.

First acute stage

The first acute stage of radiation sickness appears soon after exposure, and is characterised by temporary reversible physiological effects, whose duration (1–5 d) and severity depend on the dose;

with low doses it may not appear at all. The symptoms are typical of those seen in emotional shock, including headache, debility etc., indicating that the CNS is involved physiologically. In this respect irradiation is like any other stress, since it stimulates the general adaptation-to-stress mechanism of the body and evokes the same neuro-endocrine reaction of the pituitary-adrenal axis, though in other respects it is a very specialised stress.

Changes in cellular permeability are manifested by the release into the bloodstream of histamine and other substances (ACh, 5-HT) that are very active physiologically. Normally these substances are held inactive within the cell, and their release on irradiation provides some support for the membrane-damage hypothesis (Section 7.2.5) of the primary action of radiation. On the macro-scale, these permeability changes may be manifested by altered fluid balance, resulting in increased urine output and water intake, though this may not always appear. Decrease in plasma volume is reckoned to be a good early sign of severe radiation injury. Gastro-intestinal symptoms, again perhaps due to changes in permeability or to a nervous reaction, are nausea, vomiting and (with higher doses) diarrhoea, sometimes called the NVD *syndrome;* they are so characteristic that their severity provides a good estimate of the radiation dose.

Second chronic phase

The first stage of radiation sickness may overlap and merge into the second chronic phase, though alternatively it is possible for the early symptoms to wear off before it appears, so that the victim may congratulate himself on having recovered so quickly and easily. Even with high doses, the first stage can disappear with no indication that a fatal amount of radiation has in fact been received until the second-stage symptoms suddenly materialise.

The second stage is brought about by a *cycle of degeneration and repair* of the progenitive tissues which are the precursors of the various functional cells in the body (Figure 7.6). These precursors are actively dividing and relatively unspecialised, and are therefore more radiosensitive than the mature functional cells. Their loss after irradiation appears to be due both to direct cellular killing and to indirect toxin formation, which is detectable 2 h after a severe radiation dose. If the animal does not succumb during the first stage, it is generally able to recover, provided that there are sufficient viable precursor cells left to multiply and recolonise the depleted areas, and provided that it can remain free from infection

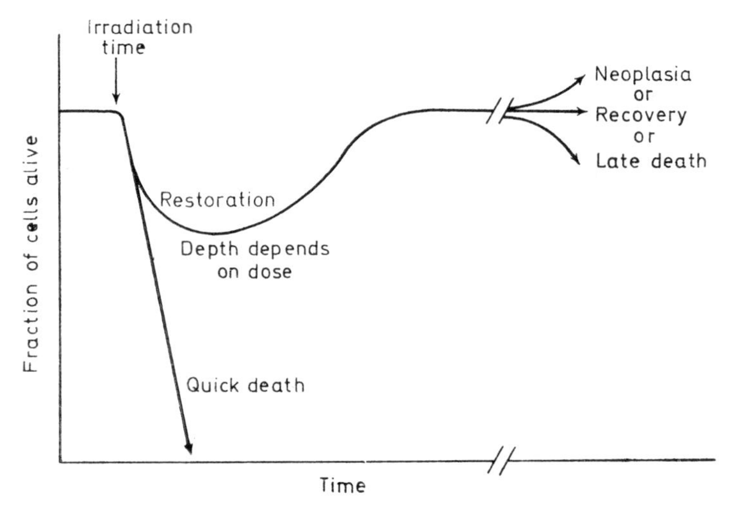

Figure 7.6. Radiation injury and recovery of progenitive tissue

(*see* below). The rate of recovery is related to the normal proliferative rate and ability of the cells, and to the quantity of viable cells left, and is reflected in the variation in general bodily fitness with time after irradiation.

Effects on different cell types

The time scale for the observed sequence of events depends on the life span of the mature functional cells; the precise nature and severity of the effects varies with the size of the dose and the dose rate, with the resistance of the organism, and with the type of radiation. Nothing is seen as long as there are enough cells left for the tissue to function normally, but as the cells die off they are not replaced by new cells (as they would be normally), their precursors having been killed, so that their numbers gradually decline; sooner or later the point is reached where there are too few cells to do the job, and the corresponding clinical symptoms then appear, sometimes quite suddenly. The several effects are shown in Table 7.2. In addition, chromosome damage may upset the reproductive machinery of the cells, and result in either cell death (failure to reproduce) or a cancerous growth (uncontrolled reproduction). All sensitive tissues will be affected by penetrating radiations, x-, gamma, and neutrons; whereas beta particles can reach only the surface tissues and so will produce predominantly conjunctivitis of the eye and skin burns.

Table 7.2. EFFECTS ON VARIOUS TYPES OF CELLS OF IRRADIATION

Cell type	*Normal life*	*Effect following irradiation*
Lymphocytes	1 day	Lowered immunity
Gut epithelium	2 days	Denudation of the gut surface; vomiting and diarrhoea
Granulocytes	7–9 days	Reduced phagocytosis: bacteria not killed
Platelets	4–5 days	Failure of blood to clot: haemorrhages
Red blood cells (RBC)	2–3 months	Reduced oxygen transport, anaemia; irradiated RBC destroyed faster
Spermatogonia	7–8 weeks	Temporary or permanent sterility
Ovaries		If all ova killed, permanent sterility
Skin	2–3 weeks	Dermatitis, ulcers, hair loss
Lens of the eye	0·5–3 years	Cataract

The lymphocytes, one type of white blood cell, are very sensitive to interphase death (Section 7.3.5); in addition, they are formed in the lymphatic tissue, which is distributed throughout the body, so that any irradiation, be it local or whole body, will affect some lymphocytes. They are therefore a very sensitive indicator of radiation exposure: changes can be observed within 2–3 h after as little as 25 rad. The granulocytes, the other type of white blood cell, are not so sensitive, their disappearance being due to death of their precursor cells, and they also provide an important indication of the severity of radiation injury. Their numbers begin to fall a day after a dose of 50–75 rad, and take at least 2–3 weeks to recover; platelets behave similarly. After a heavy dose, immature granulocytes, reticulocytes, appear in the circulation.

Skin effects

Skin effects are fairly well defined; they seem to be localised and independent of the rest of the body. As the dose rate decreases, the total dose needed for a given effect increases. Hair is temporarily lost after lower doses, 200–300 rad, and permanent loss, along with depigmentation and sparser and weaker growth, follows higher doses, of the order of 500 rad. Long-term effects are serious, involving stiffening and deformity; the first warning is alteration in the fingerprint pattern. A specialised surface tissue is the eye lens, and it is similar to skin in that the effect, the induction of cataract, is dose-rate dependent, and irradiation of the eye only is required. It has a long latent period, 5–10 years. Densely ionising radiation is particularly important in cataract, and is exceptionally more effective here than it is on other tissues.

Sterility

Induction of permanent sterility in the male dog requires practically the whole-body lethal dose, and even half of this gives no permanent sterilising effect; temporary sterility appears a few weeks after about 400 rad. The germ cells are much less resistant than the mature sperm, and their recovery from radiation is slow. On the other hand, as little as 3 rad/week induces a temporary reduction in number, viability, and motility of sperm for as long as irradiation continues, and this is one of the most sensitive indicators of radiation over-exposure. The ovaries are more sensitive than the testes, and there is wide species variation; permanent sterility needs somewhat lower doses in females than in males.

Increased susceptibility to infection

Perhaps the most serious effect in radiation sickness is increased susceptibility to infection: bacteria can enter the body more easily, and they find its defences greatly weakened when they get in. Skin burns and ulceration and the denudation of the gut remove the natural barriers against the entry of micro-organisms into the body, and even a slight wound can easily become serious. The antibody-forming ability of the body is lost with the death of the lymphocytes and, in the absence of granulocytes, invading bacteria are not destroyed. Increased capillary permeability and vascular fragility lead to escape of blood; since the platelet count is low, it does not clot, and the resultant haemorrhages bring on anaemia. Longer term anaemia also arises from damage to the erythrocyte precursor cells. And, as if all this were not enough, the body is in poor nutritional shape—no state in which to fight infection—due to its gastro-intestinal lesions; appetite is lost, and forced feeding is no use, as the damaged gut membranes can take only a small amount, if any, of even predigested food. This also has the side effect that body weight is lost. It should be emphasised that it is not the actual radiation injury so much as this great risk of secondary infection that makes chronic radiation sickness so dangerous.

Treatments

A number of treatments are available for victims of radiation sickness, antibiotics to prevent infection, electrolyte balance therapy, blood transfusions, and bone-marrow transplants (Section 7.5.6), which can give even patients who have received a normally lethal dose quite a good chance of survival.

7.4.4 Long-term radiation injuries

These injuries comprise genetic effects, leukaemia and other malignant diseases, cataracts, skin damage, impaired fertility, and ageing, i.e. premature death not attributable to specific causes. They may well all arise from the same thing, the production of point mutations, although there are probably several different mechanisms by which ionising radiation induces mutations. The end-result will differ according to the exact position of the mutation. Damage to the developing embryo (Section 7.4.5) may also be regarded as being somewhat in the long term, though it shows more quickly than some of the other effects mentioned, which may take years to appear.

Several questions are as yet unanswered, including the form of the dose–response curve, the effect of dose rate, etc.; they have to be considered separately for each kind of late radiation injury, and there is no general answer, except that there is no known threshold dose.

Genetic effects (mutations)

Genetic effects, i.e. mutations, are very important, since they persist for many years or even for generations, and in addition the gonads are very sensitive to radiation. The mutations induced by radiation are no different qualitatively from any others and, as in many other fields, radiation does not create a new phenomenon, but merely increases the rate of an existing process. The probability of a mutation occurring increases in proportion to the total cumulative radiation dose received. It has been estimated that about 2% of live births have a detrimental recessive mutation, induced by radiation or otherwise. The effect on the cell depends on how vital the affected area is to cell survival. A mutation bringing about continuous uncontrolled cell division causes a cancer to develop. Extensive chromosome defects are visible anatomically but they do not correlate with changes in function or division; profound or fatal changes may have occurred in cells that still look quite normal. Most mutations are harmful and some are lethal, as is only to be expected, since random interference with a machine as complex as the cell is likely to do far more harm than good, on the spanner-in-the-works analogy. Occasionally, however, the spanner may land in the right place to bring about a favourable mutation that provides an advantage towards survival; such mutations will persist and spread, and are an essential part of the evolutionary process. In this connection, radiation has been used as a tool in plant breeding to increase the mutation rate, and has resulted in the development of, for example, strains of rust-resistant short-stemmed barley that can be cultivated successfully where the conventional varieties cannot.

Genetic changes are not always apparent for, although every cell carries a full complement of genetic information, not all its genes are 'switched on' at one time, and indeed some are permanently 'off'. A lesion will impair cell function only if it affects a switched-on gene, or one that comes into use subsequently, and hence damage may become apparent only after some time or after the cell has divided. Recessive mutations, even if lethal, are effective only if the corresponding normal and dominant gene is absent, as may happen after cell division. In addition, a change must be observable in some way; there are undoubtedly genetic changes that exist but remain as yet undiscovered. Their revelation by the development of new techniques (e.g. developments in biochemistry that led to the discovery of various inborn metabolic errors) might be interpreted as an increase in mutation rate, but it is important to distinguish this apparent increase and the real increase resulting from greater radiation exposure or other causes. Wolff's review of radiation genetics (1967) states that radiation genetics suffers from the lack of a precise definition of the radiation-induced mutagenic lesion, though recent work has amplified the understanding of the processes involved in the formation and repair of the damage induced by ionising radiations.

The *genetic load* represents the recessive gene mutations carried by the human population. Present-day (civilised?) Man can control his environment, if somewhat precariously, and has removed the pressures of natural selection from his population, so that he can carry an increased genetic load, although there is an upper limit on how much the load can be increased before he cannot cope with the resulting defects. The deciding factor is the total dose received by the population as a whole, rather than by occupationally exposed groups which form only a small proportion of the population.

The natural genetic load arises from background radiation, which, at about 100 mrad/year, and taking the reproductive lifetime as 50 years, gives about 5 rad per generation. Another 5 rad, i.e. an extra dose of radiation at the same rate as background, is unlikely to be genetically serious, but the question is, how much more can be tolerated without undue harm? Doubling the natural mutation rate is regarded as being the maximum acceptable increase; the dose required to do this is the *doubling dose* (MRC, 1960), though the term carries a definite meaning only when the several variables, e.g. stage of development of the organism, type and dose rate of the radiation, etc., are specified. It is 30 rad in mice and fruit-flies at high dose rates, and 200 rad at lower dose rates. How much is needed in Man is not certain, but it is not less than 10 rad, and 15–30 rad has been suggested for a single exposure and 100 rad or more for chronic irradiation. Thus some extra radiation over ordinary background can be tolerated, but it is only about one order of magnitude greater; the safety factor is none too large and much more work is needed to clarify the picture and to define 'safe' limits.

Mutation is not a property restricted exclusively to germ cells, so it might appear that somatic mutations should be much more common than in fact they are. The reason why they are not is the homeostatic correcting factor of population competition: an abnormal cell is vastly outnumbered by the surrounding normal ones, and cannot survive unless it has powerfully favourable biochemical equipment, or unless the radiation kills some of the normal cells; if their poulation is depressed, the competition may be reduced sufficiently for the abnormal cell to have a better chance of survival.

Cancer

Cancer is the paradox in the action of ionising radiations, for they can both cause and cure it. This is a consequence of the duality of their action on the cell in that they can either kill it or modify its growth; the cure is brought about by an immediate lethal action, whereas the cause is a subtle delayed-action alteration leading to uncontrolled growth. It may be that radiation can cause a cell to become malignant, and/or provide an environment, maybe through the weakening of surrounding normal cells, in which a malignant cell, possibly pre-existing, can establish itself and start to grow. An alternative theory proposed more recently is that radiation may induce the formation of a carcinogenic virus.

The site where the radiation energy is absorbed naturally determines the location of the tumour; beta radiation from an external source affects the skin and the eyes, and internal bone-seeking radio-isotopes induce cancers associated with the bones. Penetrating whole-body radiation may affect all parts, producing, for example, leukaemia, although this is virtually eliminated by effective shielding of a small part of the bone marrow (Section 7.5.6). The incidence of leukaemia increases with the dose rate. The induction of cancers by local irradiation appears to have a threshold, and to require higher doses and possibly a much longer induction period than induction by non-lethal whole-body irradiation, for which there appears to be no threshold.

Leukaemia has generally been the most-discussed malignant disease, but others are becoming increasingly important as the length of time since exposure increases, because some cancers have very long latent periods, such as that of the thyroid. In survivors of the 1945 Japanese atomic bomb explosions, the combined total of cancers other than leukaemia now appears to be greater than the number of leukaemias, and accumulating information from them and from other groups exposed a long time ago suggests that, after uniform whole-body exposure, other cancers will ultimately be an order of magnitude more frequent than leukaemia.

Ageing

Ageing (Lindop, 1966) is seen as an acceleration of senescence; vitality and resistance to infection are lowered, and mental and physical fitness deteriorate. It is not known if radiation accelerates normal ageing processes or introduces a new factor, nor how much of the decrease in life expectancy is due to premature ageing and how much to increased incidence of radiation-induced diseases. Increased incidence of somatic mutations may be involved as a possible mechanism. Experimentation is difficult because the animal has to be observed for the whole of its life span, and large groups are necessary since life spans vary anyway. Information on large animals is therefore scarce and, as small animals may not react in the same manner, results should not be extrapolated. There may be a tolerance dose below which there is no shortening of life; no ill effect has yet been shown to result from a dose of 100 rad spread over several years. Data from radiologists and other users of radiation are contradictory; the life expectation of a group that had received 1000 rad over the working life was found to be reduced by 5 years, but another similar group was shown to be unaffected. There is indeed evidence from experiments on flour-beetles and *Paramecium* that very low doses of radiation may actually increase life span, so that radiation, like many drugs, may have physiological effects at low concentrations that shade into pharmacological effects as the concentration is increased.

Mole (1970) raises an objection to the term 'ageing'. All the other late effects of radiation are definite kinds of damage, but ageing is non-specific, is not a medical diagnosis, and would not be acceptable on a death certificate as a description of the cause of death. After 25 years of intensive investigation of irradiated humans and animals, it seems most unlikely that there will be any kinds of radiation effect that have not been anticipated or observed, and it is surely a retrograde step to describe the increase in the death rate from radiation in terms of some non-specific concept. The question is further discussed in ICRP (1969).

Risks

'What are the risks?' is the question to be asked of long-term radiation injuries, above all others, since they are taken to be the most serious effects of radiation (Section 7.4.1). The risk (ICRP, 1967) is assessed by comparison of the incidence of an injury from man-made radiation with its incidence from natural background radiation and from other causes. Since some uncertainty is always attached to experimental results (Section 7.4.1), and they are statistically too unreliable to be quoted exactly, it is better to regard them only as estimates, involving a certain degree of guess-work, of the *order of risk* from the radiation, e.g. 10–100 incidences of a particular injury per 10^6 individuals exposed, an occurrence in the range 1×10^{-5} to 10×10^{-5}, is a 5 th-order risk. Exposure to 1 rad is estimated to carry a 5th-order risk of cancer, though this is by extrapolation from high-dose experiments. The natural mutation rate is about 250–300 per 10^6 and, assuming that it would be doubled by a single dose of 20 rad, 1 rad would therefore give 12–15 mutations per 10^6, a 5th-order risk of genetic damage. The occurrence of chromosome aberrations is high naturally, 2nd order in stillbirths and 3rd order in live births, but the radiation effect cannot be reliably estimated since it is not fully understood how radiation acts on chromosomes.

It was claimed recently that radiation from strontium-90 fallout following weapons tests significantly increases the risk of genetic damage, and that it has injured or killed 400 000 children in the United States of America and 80 000 in Britain since 1945 (Sternglass, 1969). However, the claim does not stand up to careful analysis (Lindop and Rotblat, 1969), and an expert committee of the Medical Research Council has concluded that the evidence does not prove or even support it.

7.4.5 Irradiation and the developing organism

The developing embryo is considerably more sensitive to radiation than the adult, since its tissues are actively dividing, and at certain stages in addition they are differentiating: relatively small doses therefore interfere with embryonic development, and as little as 25–50 rad may be effective. For the same reason, the effects are completely different from anything that can be produced by irradiation of the adult. The primary intracellular lesion, chromosome breakage, leads to the initial cellular effect, cell death and, although ionisation is distributed evenly throughout the embryo, the cells dying are probably not. The pattern of cellular descent is thus interrupted, leading directly or indirectly to the observed damage or developmental effect (Russell and Russell, 1954).

The proportion of early mammalian embryos killed by irradiation is initially high, and gradually declines; those surviving can overcome the radiation damage and develop more or less normally, and very few have any visible abnormality. This corresponds to the stage of development when there is no apparent cellular differentiation. At the point when the three primary germinal layers form (the eighth day of gestation in the rat and the sixth in the mouse) the incidence of abnormalities in all embryos suddenly shoots up to 100%, and remains high for the duration of major organogenesis. In addition, many embryos are so badly damaged that, though able to exist *in utero*, they die at or immediately after birth. The organ system (or systems) affected depends on which is developing at the time of irradiation; in mammals the nervous and circulatory systems develop first, followed by the sense organs, skeleton, muscles and viscera (Wilson, 1954), and a similar picture is seen in the developing amphibian (Rugh, 1954).

Once organogenesis is complete, the embryo merely increases in size, with little or no further development. Irradiation thereafter with the same doses as before produces neither death nor abnormalities of form, but will restrict growth, resulting in a stunted though topographically normal foetus. Physiological symptoms have also been recorded.

The most vulnerable stage in Man is the second to the seventh week of gestation, i.e. so early that pregnancy may not be suspected, and for this reason it has been recommanded that pelvic irradiation of women of child-bearing age should be restricted to the fortnight immediately following menstruation to avoid unwittingly irradiating an early foetus.

7.4.6 Comparative radiosensitivities of organisms

A very wide range of sensitivities has been recorded, and some types of organisms can withstand, apparently unharmed, doses that kill others. Even within the same species, there are variations from strain to strain, hybrids being generally more resistant than pure breeds, and there may be large individual variations. It is not yet possible to explain all these facts fully, since complete survival data are not available; in addition, the results of different workers are not always comparable due to wide differences in experimental technique, and in the criteria used to measure 'sensitivity'. However, a number of general conclusions emerge.

Unicellular organisms are very resistant, with a few exceptions, and there is a trend towards increasing sensitivity as the organism becomes more complex, but wide variations may be found among similar organisms. This may be because the cells of more complex organisms contain more DNA; size of the nucleus is linked to its DNA content, and an inverse relation has been found between nuclear volume and radioresistance among several species of plants. Table 7.3 shows lethal doses for various organisms.

Insects are relatively resistant among highly developed animals, the adults being more resistant than the young: the lethal dose is about 100 times greater than in mammals. This may be because all their cells except those in the sex organs have ceased to divide; alternatively, their tissue oxygen tension may be low as a result of tracheal respiration and their radiosensitivity correspondingly reduced (Section 7.5.2). Like most invertebrates, they maintain their osmotic balance largely by amino-acids or small polypeptides, which are known to give some protection (Section 7.5.5).

Mammals are more sensitive than other vertebrates; this is easily explained in comparison with poikilotherms, whose body temperaures are lower, but not so with birds, whose body temperatures are slightly higher than in mammals, e.g. pigeons are about five times as resistant as rats, and the question is, why does a bird resist?

The sensitivity of plants increases from monocotyledons through angiosperms to gymnosperms: the growth of pine trees is inhibited by 2 rad/d and they are killed within a year by 4000 rad cumulative dose, whereas sedges withstand up to 350 rad/d. Radiosensitivity of sunflower seedlings was markedly decreased by withholding certain essential elements (B, Ca, K) for 3 d before irradiation, the treatment reducing growth rate and cellular activity (Wiecek and Skok, 1968).

Table 7.3. RADIOSENSITIVITY OF VARIOUS ORGANISMS EXPRESSED AS LD_{50} (30 d) EXCEPT WHERE STATED OTHERWISE, e.g. (5) MEANS LD_{50} (5 d)

Organism	*Radiosensitivity*
Micro-organisms	(krad)
Paramecium	300
Amoeba	100
Yeast	30
Escherischia coli	5·6
Invertebrates	(krad)
Radix japonica (mollusc)	2(80)
	12(20)
Thais (Nucella) (dog-whelk)	17(80)
	20(5)
Artemia young (brine-shrimp)	80(5)
	20(10)
Chaorborus (phantom-gnat) larvae	10–25
Snail	8–20
Vertebrates	(rad)
Goldfish	670
Frog	700
Newt	3000
Tortoise	1500
Chicken	600–800
Guinea-pig	250
Pig	600 (gamma), 350–400 (x-rays)
Various adult rodents (mouse, rat, hamster, rabbit)	550–700 (approx.)
Dog	335
Goat	350
Man, monkey	300

The age of the organism affects its radiosensitivity: in mammals the LD_{50} (30 d) increases from birth to adulthood, then falls again in old age, and the same pattern is seen in the fruit-fly *Drosophila*. The number of chromosomes is important, comparison of diploid and haploid strains of yeasts and of chrysanthemum plants of varying degrees of polyploidy having shown that radioresistance increases with the ploidy. It might be that the polyploids have some 'spare' chromosomes, as it were, so that damage to one is not as catastrophic as in the haploid with its one set of genetic material. The liver of adult rodents is polyploid, to some extent, which might account for their resistance being greater than that of the young animal. Other hypotheses to explain differences in radiosensitivity are that enzyme systems sensitive to cyanide (those containing heavy metals) are the most radiosensitive—some insects are able to withstand cyanide—and that the dehydrogenase content of a tissue might be related to its resistance.

7.5. MODIFICATION OF RADIATION EFFECTS

7.5.1 Restoration after radiation injury

The discussion of recovery from radiation injury and how the damage is repaired is hampered by the areas of ignorance and uncertainty about the biological action of ionising radiations. Little is known for certain about the nature of the initial lesion or the mechanism through which it is translated into the observed effects, though it is unlikely that one single simple system is operating. In addition, extrapolation from the simplest elements of the cell to the whole organism is difficult or impossible. More work to clarify the picture is desirable, since a precise understanding of restoration processes will provide a scientific basis for the treatment of radiation victims (Section 7.4.3).

All forms of life, from micro-organisms to mammals, can recover from radiation damage. The existence of recovery mechanisms is shown by the sparing effect of the *split-dose technique*, i.e. two irradiations spaced out do less damage than the same total dose given all at once. Techniques for the study of early radiation effects, in particular reversible physiological symptoms, are very well suited for the measurement of repair and recovery processes, e.g. the emission of light from luminous bacteria and the uptake of ions by plant roots (Section 7.3.2). These (and other) experiments also show that there are two kinds of damage, one repairable and the other not. The latter is therefore cumulative and irreversible, and often long term and/or delayed action; it need not necessarily show up in the short term. Thus mammals can recover from short-term radiation sickness, but long-term effects are permanent. The two kinds of damage can be distinguished by their different reactions to change in dose rate: the extent of repairable damage becomes less as the dose rate is lowered, whereas permanent damage is unaffected. Repair processes can be considered to be an abnormal requirement of the cell, differing from its usual synthetic processes; it has been found that erythropoiesis in irradiated mice is different from normal.

In theory at least, though facts are not always easy to reconcile with theories, there are three possibilities for restoration processes: first, *protection*, the neutralisation of primary effects; secondly, *intracellular repair* of secondary effects by re-synthesis of destroyed or inactivated molecules, which restore sublethal injury; and thirdly, *recovery* of the organ or tissue by proliferation of cells to replace those that have been killed. The two last have been shown to be general restoration mechanisms in mammalian cells. Cells can survive if the damage is not too extensive; the split-dose technique has demonstrated that repair starts immeditely after irradiation and, within minutes or hours, or a very few days at the most, the greater part, if not all, of it has been carried out. Damaged cells can also be selectively eliminated at division; it has to be assumed that mutated cells are less likely to survive, although long-term effects may arise from mutations which give a selective advantage (or at least no disadvantage) for survival. Chromosome breakage is an example of repairable damage; most of it is repaired soon after irradiation, and the rest is eliminated either by a slow repair process that finally makes the cell look normal, or by cell selection at division. On the other hand, point mutations, be they spontaneous or radiation induced, are not repairable. They are likely to be responsible for delayed radiation effects and the degeneration with time of the cell renewal systems. As they are not repaired and are not eliminated by cell selection, point mutations gradually accumulate and finally result in either degeneration or a neoplastic response (Brookhaven, 1967).

It has been found empirically that, though fractionation of the dose, i.e. giving it in several parts rather than all at once, increases the total amount of radiation needed to cure a cancer, it improves the differential sensitivity between tumour and healthy tissue. It appears that the cancer recovers more slowly than the normal cells between the fractions of the dose and thereby suffers more damage overall.

7.5.2 The oxygen effect

This effect is a general phenomenon in radiobiology. It refers to the function of oxygen in the formation of the initial lesion, and it should not be confused with two other facts: first that oxygen itself can damage cells without any radiation at all, if its partial pressure is too high (above that in air), and secondly that post-irradiation development of the lesion, as distinct from its formation, requires metabolism (Section 7.2.5) and therefore oxygen.

If the oxygen concentration of cells or tissues is reduced sufficiently, the damage done by x- and gamma-radiation is diminished, though not abolished altogether, i.e. hypoxia or anoxia increases radioresistance. For the protective effect to be seen, the tissues must be anoxic or hypoxic during the period of irradiation, when the initial chemical changes are taking place, and exposure to oxygen before or after is immaterial provided that it is absent during this crucial period; indeed, the radiosensitivity of bacteria has been shown to follow changes in oxygenation within a fraction of a second. If therefore seems likely that oxygen modifies the primary radiation lesion. A possible explanation is that, though the formation of an organic free radical R· (Section 7.2.4) may be reversed in the absence of oxygen by combination with a hydrogen radical and no damage result ($R\cdot + H\cdot \rightarrow RH$), in the presence of oxygen, damaging peroxy-radicals may be formed and the organic molecule irreversibly altered:

$$R\cdot + O_2 \rightarrow RO_2\cdot \text{ and } H\cdot + O_2 \rightarrow HO_2\cdot$$

Free-radical chain reactions may also occur with oxygen. Another possibility is that 'fixation' of the radiation lesion takes place, by electron-affinic oxygen removing the electron produced in the initial ionising event. An explanation which is not now believed to be true is that, under hypoxic conditions in a mammal, reflex stimulation of the sympathetic nervous system releases large amounts of adrenaline and noradrenaline, which are phenolic amines and therefore good radioprotectors (Section 7.5.5). It may be noted that the oxygen effect can be reversed by inert gases, whose pressure to be effective is inversely proportional to their oil solubility, and (in beans at least) this appears to be the result of inhibiting the diffusion of oxygen into radiosensitive cells via the intracellular spaces. Densely ionising radiations (alphas, neutrons) are much less affected by the presence or absence of oxygen than the sparsely-ionising x- and gamma-radiations; the same is true of chemical protection (Section 7.5.5). This may be because densely ionising radiations cause several ionisations within one macromolecule and damage it too much for it to be restored, even in anoxic conditions; the chemical lesion here needs no enhancement by oxygen. It is also found that the effect of chemical protectors is much more evident in the presence of oxygen.

The change from maximal to minimal radioresistance takes place over a narrow range of oxygen tension, from zero to a few torr (1 torr = 133 N/m^2); there is typically a threefold change in sensitivity. A useful parameter is the oxygen tension at which half the total increase in sensitivity is attained; it has been found to be about 2·5–5 torr for various cells. This quick change of sensitivity over a small range means that the exact oxygen tension in cells and tissues is very important, but unfortunately its measurement is very difficult and it has been said that sensitivity depends more on the technique used to measure oxygen tension that on the actual measurement. In a cell culture, the intra-cellular oxygen concentration approximates to that in the surrounding medium, but will not necessarily be identical with it, especially if the cells are respiring quickly. Mammalian tissues have about the same oxygen tension as the outflowing venous blood, 20–25 torr. Though complete deoxygenation of a solution may be a slow process—reduction of dissolved oxygen to 2 μmol/l by flushing with nitrogen can take up to 5 h—cells and tissues *in vivo* can change from oxygenated to anoxic conditions very quickly.

The oxygen effect can be relevant in cancer radiotherapy. Poorly vascularised tumours are likely to be low in oxygen; if the blood flow to them can be increased by some means (e.g. vasodilatory drugs) they will receive more oxygen and thereby become more sensitive. The surrounding healthy tissue will also receive more blood, of course but, since it is already well oxygenated, its sensitivity will already be maximal and will not be increased by further oxygen. Alternatively, the patient might be made completely anoxic, to reduce the sensitivity of healthy tissue, but lack of oxygen quickly damages brain cells so this approach is hardly practicable.

7.5.3 Hydration

Since most of the damaging effects of radiation take place through free radicals formed from water, it is to be expected that dry systems would be in general less sensitive than those containing water, since fewer free radicals are formed. However, free radicals may be trapped in a dry system and exist for a comparatively long time, in contrast to their transient existence in water. Hydration also influences the physiological factors which develop the radiation injury. Dry seeds of maize and barley were damaged less by soaking in D_2O than in H_2O before irradiation; the effect is largely oxygen-dependent (Gaur, Joshi, and Notani, 1969).

Work on dehydration is necessarily limited to micro-organisms and dry seeds of plants, and is not applicable to the radiobiology of higher animals.

7.5.4 Temperature

Temperature has a complex effect, which is not yet fully clarified. It is known that bacterial cultures show a sudden sharp increase in resistance just below freezing-point, maybe because the diffusion of free radicals is hindered in ice, though it is questionable whether intracellular water actually freezes in the common sense of the word. It may prove possible to treat cancers by cooling the rest of the body (for which the technology already exists) and warming the tumour locally during its irradiation.

7.5.5 Chemical protection

The attack of free radicals on biological molecules is essentially a chemical process. Since it depends on the reactivity, concentration, and size of the molecule (Section 7.2.3), the addition of another compound may influence the extent to which radicals react with the first, which may thus be either protected from or sensitised to attack. The best radioprotectors are compounds that can effectively compete for free radicals with the biological target molecules, or react more readily with them. Chemical protection is defined as the addition before irradiation of a chemical substance which lessens radiation damage, in contrast to therapeutic measures that seek to ameliorate radiation effects by chemical or drug treatment after irradiation. As with the oxygen effect, chemical protection against alphas and neutrons is much weaker than against x-rays.

How radioprotectors act is not known for certain, though obviously any theory must agree with *in vivo* and *in vitro* observations. There may well be several possible mechanisms, acting simultaneously and synergistically, and their relative importance may vary with the protector, the tissue, oxygen tension, the quality of the radiation, and other factors. Temperature seems important, though more work is needed. By definition, and as proved by experiment, the protector must be present during irradiation, so that it almost certainly acts at the stage of formation of the initial chemical lesion, confirmation of which is that any parameter of radiation damage can be taken as an indicator of protection. To explain how a protector acts, the mechanism of formation of the lesion should therefore be known. Three possibilities (among others) have been put forward as the basis of protector action:

1. Energy is transferred to the protector so that, although energy may be deposited in a biological target molecule, its chemical effect is on the protector.
2. The lesion is at first reversible, i.e. repairable, and by reaction with the protector the molecule is restored to its original state. In the absence of the protector it is irreversibly altered.
3. The physiological state of the system is changed through a pharmacological action of the protector.

Many different substances are protective. One of the most effective yet found is the amino-acid cysteine and its decarboxylated form cysteamine (β-mercapto-ethylamine or MEA): a pre-irradiation dose of 3 mg of the latter protects mice completely from the otherwise lethal effects of 700 rad of x-rays. Studies of its analogues show that the active structure is a free or potential sulphydryl (—SH) group at one end of a two- or three-carbon chain, and a strongly basic group, amine or guanidine, at the other end. An example is a thiourea derivative, AET (β-amino-ethyl-iso-thiourea) which is converted in water (and in the body) to MEG (mercapto-ethyl guanidine), the active form.

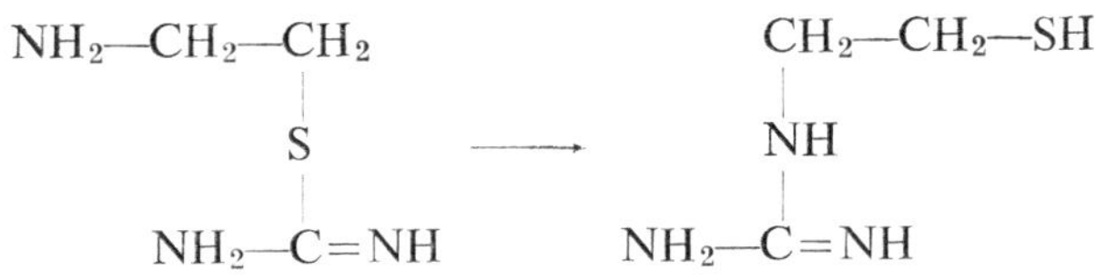

$$\begin{array}{ccc}
NH_2\text{—}CH_2\text{—}CH_2 & & CH_2\text{—}CH_2\text{—}SH \\
\qquad\qquad\quad | & & | \qquad\qquad\quad \\
\qquad\qquad\quad S & \longrightarrow & NH \qquad\qquad\quad \\
\qquad\qquad\quad | & & | \qquad\qquad\quad \\
NH_2\text{—}C\text{=}NH & & NH_2\text{—}C\text{=}NH
\end{array}$$

The protection given by these compounds may depend on their intracellular concentration, as it falls off in step with their excretion; this might also account for their efficiency as radioprotectors, since they are concentrated by several tissues, including one which is critical in irradiation, the bone marrow. Their protective effect is increased by incubation before irradiation. The sulphhydryl group on its own is insufficient to protect, and in fact many non-physiological —SH compounds are inactive or are even sensitisers.

Sulphur-containing radioprotectors are fairly reactive compounds and have been suggested to act by some or all of at least three mechanisms. They can lower intracellular oxygen concentration; this alone cannot explain experimental evidence, but it may be a contributory factor. Their action is not due to their reducing power. They might compete for, or 'scavenge' free radicals, or repair the damage they cause, but rather too high concentrations are required to achieve this than are effective *in vivo*. They can form mixed disulphides, linking an —SH group in a protein with one in the protector, instead of the usual all-protein disulphide bridge (Modig, 1968), though there is no correlation between protection and the binding of sulphur compounds in this way.

The latest hypothesis (Brookhaven, 1967) is that sulphur protectors induce a profound biochemical shock, the three disturbances being: first, a very rapid activation of the hypothalamic-pituitary-adrenal neuro-endocrine systems; secondly, cardiovascular disturbances (though these correlate very little with protection); and thirdly, inhibition of carbohydrate metabolism leading to a decrease in oxygen consumption and respiratory quotient, which correlate very well in time with the intensity of protection and might link with rapid and reversible ultrastructural changes in the mitochondria. By delaying mitosis and DNA synthesis, they also protect against genetic damage. It is suggested that these biochemical changes increase the efficiency of the repair system in the cell or protect it from radiation damage and, though the theory is not proven, there is as yet nothing to contradict it.

Subcutaneous injection of cysteamine has been shown in mice to be effective only in the area of the injection, and without influence on the rest of the body. It can therefore provide *local protection*, in contrast to its relative cystamine and to histamine, which can exert pharmacological action over the whole body from subcutaneous injection at one point.

Cyanides or nitriles in non-lethal (but still toxic) doses are weakly protective; as are also amino acids and peptides, so that all proteins within the cell to some extent protect each other. Amines, especially if aromatic, are much more active. They include 5-hydroxy-tryptamine (5-HT or serotonin), adrenaline, noradrenaline, and histamine, which occur naturally in many animals. Histamine is vasodilatory and slows the circulation, so that it may act by reducing oxygen tension in the tissues; other amines are thought to act similarly by induction of tissue hypoxia. A wide variety of other substances, including dimethyl sulphoxide (Ashwood Smith, 1967), sex hormones, anoxic agents, some anaesthetics and analgesics, and hydroxy-compounds, are also weakly active; chelating agents are very effective but unfortunately are very toxic due to interference with calcium. Some substances act slowly, needing to be administered several hours before irradiation (e.g. parathyroid extract) and it is likely that they induce a physiological change that lessens sensitivity. Mixtures of protectors may be useful, and they may have a synergistic effect on each other (Maisin *et al.*, 1968). Foye (1969) reviews radioprotective agents in mammals, and Pennington and Meloan (1968) deal with those containing sulphur.

Sensitisers

An equally wide variety of substances are sensitisers, intensifying the effects of radiation. They include the phosphate ester of a quinone compound related to vitamin K, known as Synkavit, actinomycin D, iodo-acetate, tetracyclines and 5-bromo- and 5-iodo-deoxy-uridine. Their mode of action is unknown, but might be due to interference with normally present physiological protective agents.

7.5.6 Post-irradiation therapy

Chemical protection offers one approach towards lessening the effects of radiation, but is limited in use. How it works is, as just seen, not fully understood, and even less so is post-irradiation chemotherapy, which seeks to repair, rather than to prevent, radiation injury or to stimulate natural recovery processes. Some drugs can enhance recovery, e.g. geranyl hydroquinone: though without effect on the mortality of irradiated mice, it accelerates restoration of white blood cells in the survivors.

More is known about *treatment with bone marrow and spleen cells*. Local irradiation of only a part of the body can be far more intense than whole-body irradiation without causing harm, and whole-body irradiation is ameliorated by local shielding of the limbs or of the spleen; shielding of other organs is much less effective (e.g. Fujioka *et al.*, 1967). This arises from the influence of the haematopoietic cells in the shielded regions; unlike chemical protectors, they act not during irradiation but during the following hours. The destroyed tissues are repopulated by cells conveyed in the blood from similar undamaged tissues, showing that 'metastasis' is not restricted to cancers.

Treatment with these cells enables animals which have received a normally fatal dose to survive, spleen homogenates but not extracts being effective. The effectiveness decreases as the relationship between donor and recipient weakens, i.e. autologous > isologous > homologous > heterologous. Two theories were developed to explain the effect: the humoral theory proposed that an unstable factor from the injected material (homogenate or cells) stimulated regeneration of the depleted areas; the cellular theory, that cells were being grafted, explained why spleen extracts did not work, but had one great stumbling block, that heterologous (and even homologous) tissue grafts are known not to 'take', and are rejected by an immune reaction of the host. Unlikely though it once seemed, the cellular theory has now been proved to be true, but analysis of the treatment shows that heterologous grafts only take if a fatal (LD_{100}) dose is given over the whole body. If the animals receive smaller doses, even those making them very sick but able to recover, the graft can kill them, for the immune response is not completely destroyed, as it is with the fatal dose, and the additional stress from the injection of foreign material proves too much for the animal. Heterologous bone-marrow or spleen therapy is therefore hazardous if accurate dosimetry is not available. A further complication is secondary disease, if the host regenerates his immune reaction and eliminates the graft, though his own haematopoietic tissue may by then have recovered sufficiently to enable him to survive. Sometimes the graft can develop antibodies against the host, a very serious and dangerous complication. A chimaera is a mythical beast made up of parts of two or more animals and, since haematopoietic cell injections act by survival of cells from one animal inside another, animals so treated are aptly named radiation chimaeras. A book by Van Bekkum and De Vries (1967) tells the story of their history and development.

REFERENCES

ALEXANDER, P., (1965). *Atomic Radiation and Life*, 2nd edn, Pelican No. A399, Penguin, Harmondsworth, 296 pp.

ASHWOOD SMITH, M. J., (1967). 'Radioprotective Properties of DMSO in Cellular Systems', *Ann. N.Y. Acad. Sci.*, **141,** 45–62.

BACQ, Z. M., and ALEXANDER, P., (1961). *Fundamentals of Radiobiology*, Pergamon, Oxford, 555 pp.

BROOKHAVEN, (1967). 'Recovery and Repair Mechanisms in Radiobiology', *Brookhaven Symp. Biol.*, No. 20, 316 pp.

CASARETT, A. P., (1968). *Radiation Biology*, Prentice-Hall, N.J., 368 pp.

ELKIND, M. M., and WHITMORE, G. F., (1968). *The Radiobiology of Cultured Mammalian Cells*, Gordon and Breach, New York.

FISCHINGER, P. J., and O'CONNOR, T. E., (1969). 'Radiation Leukaemia Virus', *Science, N.Y.*, **165,** 306–309.

FOYE, W. O., (1969). 'Review of Radioprotective Agents in Mammals', *Jnl. pharm. Sci.*, **58,** 283–300.

FUJIOKA, S., HIRASHIMA, K., KUMATORI, T., TAKAKU, F., and NAKAO, K., (1967). 'Mechanism of Haematopoietic Recovery in the X-irradiated Mouse with Spleen or One Leg Shielded', *Radiat. Res.*, **31,** 826–839.

GAUR, B. K., JOSHI, R. K., and NOTANI, N. K., (1969). 'Effect of Heavy Water on Radiosensitivity of Maize and Barley Seeds', *Radiat. Bot.*, **9,** 61–67.

ICRP, (1967). 'Evaluation of the Risks from Radiation', *ICRP Publ.*, No. 8, Pergamon, Oxford.

ICRP, (1969). 'Radiosensitivity and Spatial Distribution of Dose', *ICRP Publ.*, No. 14, Pergamon, Oxford.

KUZIN, A. M., PLYSHEVSKAYA, E. G., KOPYLOV, V. A., IVANITSKAYA, E. A., LEBEDEVA, N. E., KOLOMIITSEVA, I. K., TOKARSKAYA, V. I., and MELNIKOVA, S. K., (1966). 'The Orthophenol–Orthoquinone System in the Initial Mechanism of Action of Ionising Radiation', *Fedn. Proc.*, **25,** T675–T682.

LAMERTON, L. F., (1968). 'Radiation Biology and Cell Population Kinetics', *Physics Med. Biol.*, **13,** 1–14.

LINDOP, P. J., (1966). 'Proc. colloquium Radiat. Ageing', Ed. SACHER, G. A., Taylor and Francis, London, 456 pp.

LINDOP, P. J., and ROTBLAT, J., (1969). 'Strontium-90 and Infant Mortality', *Nature, Lond.*, **224,** 1257–1261.

LUSHBAUGH, G. C., COMOS, F., and HOFSTRA, R., (1967). 'Clinical Studies of Radiation Effects in Man', *Radiat. Res. Suppl.*, **7,** 398.

LUSHBAUGH, G. C., *et al.*, (1968). Paper 17.1 in 'Dose Rate in Mammalian Radiation Biology', UT-AEC. Conf. 680 410, United States Atomic Energy Commission Division of Technical Information.

MRC, (1960). *Hazards to Man of Nuclear and Allied Radiations*, CMND 1225, HMSO, London, 154 pp.

MAISIN, J. R., MATTELIN, G., FRIDMAN-MANDUZIO, A., and VANDERPARREN, J., (1968). 'Reduction of Short- and Long-Term Radiation Lethality by Mixtures of Protectors', *Radiat. Res.*, **35**, 26–45.

MODIG, H., (1968). 'Cellular Mixed Disuphides Between Thiols and Proteins and their Possible Implication for Radiation Protection', *Biochem. Pharmac.*, **17**, 177–186.

MOLE, R. H., (1970). 'Radiation Effects in Man: Current Views and Prospects', Paper 83, 2nd int. Congr. IRPA, 1970.

PENNINGTON, S. N., and MELOAN, C. E., (1968). 'Study of Radiation Protection by Sulphur Compounds', *Radiat. Bot.*, **8**, 345–354.

RUGH, R., (1954). 'Effect of Ionising Radiation on Amphibian Development', *J. cell comp. Physiol.*, **43**, Suppl. 1, 39–75.

RUSSELL, L. B., and RUSSELL, W. L., (1954). 'Analysis of the Changing Response of the Developing Mouse Embryo', *J. cell. comp. Physiol.*, **43**, Suppl. 1, 103–149.

STERNGLASS, E., (1969). 'Has Nuclear Testing Caused Infant Deaths?' *New Scient.*, **43**, 178–181.

STINSON, J. C., and MOORE, J. C., (1966). 'Histochemical Technique for Direct Observation of Radiation Reactions', *Radiology*, **87**, 527–528.

USAEC, (1968). *Index of Radiation Biology (24 000 references up to 1960), see Hlth Phys.*, 1968, **14**, 73.

van BEKKUM, D. W., and DE VRIES, M. J., (1967). *Radiation Chimaeras*, Academic Press, New York, 277 pp.

WIECEK, C. S., and SKOK, J., (1968). 'Effects of Brief Withholding of Essential Elements on Radiosensitivity of Sunflower Plants', *Radiat. Bot.*, **8**, 245–250.

WILSON, J. G., (1954). 'Differentiation and the Reaction of Rat Embryos to Radiation', *J. cell. comp. Physiol.*, **43**, Suppl. 1, 11–37.

WOLFF, S., (1967). 'Radiation Genetics', *A. Rev. Genet.*, **1**, 221–244.

Chapter 8

RADIOACTIVITY IN THE BIOSPHERE

8.1 FALL-OUT AND FOOD CHAINS

8.1.1 Sources of environmental contamination

The hazard from radio-isotopes in the biosphere arises through their entry into food chains and their transmission along them into Man's diet. They are therefore important as potential sources of internal irradiation, and external irradiation is unimportant since their concentrations and the external radiation doses from them are comparatively low. Radio-isotopes entering the biosphere may be either artificial or natural. Little or nothing can be done about the distribution and spread of those that occur naturally, and in any case we have learnt, as it were, to live with them, so that for this reason artificial isotopes, which are also more dangerous, cause the most concern.

By far the greatest contributor to environmental contamination up till now has been fall-out, the release of radio-isotopes into the biosphere following the detonation of a nuclear weapon or 'device' based on the fission of heavy nuclei such as uranium-235 or plutonium-239. The isotopes in fall-out are mainly *fission products;* some *induced activity* may arise from the capture by stable elements of neutrons released in the fission, but its contribution to dietary contamination is very small compared to that from fission products. It is unlikely that Man's diet will ever be completely free of fission products, though the amounts in it will steadily fall as time passes if the Nuclear Test Ban agreement is observed.

Three types of fall-out are responsible for the contamination of land, water, and air. *Local* fall-out is confined to the immediate locality of the explosion; a device going off underground produces nothing else, as it contaminates the atmosphere little or not at all, and the fission products do not therefore spread very far away from the place where they were formed. The other two types, on the other hand, do involve atmospheric contamination, and are therefore much more hazardous since they affect a much larger area of the Earth's surface. *Tropospheric* fall-out arises from relatively small atmospheric explosions; it spreads round the world in a band in the latitude of the explosion. Large atmospheric detonations greater than one megaton (the same destructive power as a million tons of TNT) produce chiefly *stratospheric* fall-out. Their mushroom cloud of radioactivity reaches up into the stratosphere, whose winds disperse it world-wide, and from which it takes a long time—6 months to 3 years or more—to fall out. Underground tests may or may not be safer than those in the atmosphere; it is thought that large ones may be able to trigger off earthquakes, and that fall-out may be able to penetrate aquifers (water-bearing strata) and thence directly enter drinking water supplies.

Among other man-made radiation sources, the medical use of radio-isotopes, and nuclear reactors, are of only minor or very localised importance in the contamination of the environment, though concern has been expressed recently about the increasing numbers of reactors and the increasing chance of an accident with serious consequences in a reactor plant, and the suggestion made that sources of nuclear power involving fusion rather than fission reactions should be developed instead. A potential new source of environmental radioactivity is the use of nuclear devices for excavation and earth-moving (Charnell, Zorich, and Holly, 1969; Martin 1969).

Waste disposal could be very dangerous, but its hazards have been appraised and its sources strictly controlled, and indeed it is a model of the marshalling of talent on an international scale to mitigate or avoid the world-wide problems that would have resulted from uncontrolled and unlimited discharge of radioactive wastes into the biosphere. There are some unsolved questions, but much has been learnt from radioactivity that is applicable to other wastes. In sharp contrast has been the development of other technological achievements resulting in environmental pollution, notably of pesticides, for there is as yet far too little realisation of the fact that there is a limit to the amount of biologically active chemicals that can be tolerated in circulation in the biosphere. The attitude has been that toxic materials will be diluted so much that they will disappear, and the corollary of this, the 'right' to pollute the environment, to dispose of anything without restriction, has been assumed. But the concept of dilution to innocuousness is false, for biologically active materials travel in well-defined patterns, and their dilution in the inanimate environment is nullified by their concentration by living things from that environment. Indeed, the presence of man-made radioactivity in the world has proved this very fact, and accumulation by organisms is a key factor in limiting the quantity of radioactive waste discharge and the amount of man-made radioactivity in the biosphere. But it is to our shame that our present-day technology has created the aptly-named effluent society, that for the most part disposes of its wastes so irresponsibly that it has created, and is steadily building up, a large biological debt that its children will have to pay in the not-so-far-distant future (Woodwell, 1969).

8.1.2 Critical factors

The critical factors in environmental contamination and its transfer to Man are distinguished as those that give the greatest radiation risk or hazard. The critical organ (Section 6.1.4) is that which suffers most damage from a particular nuclide. A critical population group is the part of a population which is at greatest risk, i.e. which is liable to suffer the most damage from radiation, in a particular set of circumstances. Critical path(s) may be distinguished as the route, or routes, through the food chain which carry most or all of the activity of a particular nuclide. Several different paths are available for the passage of nuclides from the environment to man, but not all nuclides use all routes, and sometimes only one path may be critical. At the end of a critical path are the critical foods which are eaten by Man, the main sources of radioactivity in the diet; they may also serve as *pilot* or *indicator foods*, for analysis of them alone will give a very good estimate of the total activity in the diet, and obviates the necessity for analysing everything that is eaten. An acute localised contaminating event is likely to be of greater concern in its immediate neighbourhood than general stratospheric fall-out, as it will there produce more fall-out, and it is thus a critical circumstance or critical event.

Finally, there are the critical nuclides. Fission products comprise many different isotopes of many elements; their importance is determined by their production yield and half-life, how fast and how readily they enter food chains and foods eaten by Man, the extent to which they are absorbed from the gut when he has consumed them, and how much is deposited in the critical organ and how long it is retained there. Many fission products are short-lived, or formed in small amounts, or of little biological significance, and only a few are important enough to be regarded as critical nuclides. Those of greatest, in fact world-wide importance, are the 'three ugly sisters' of radiobiology: strontium-90, caesium-137, and iodine-131. The first two have long half-lives (28 and 30 years) and are readily taken up and retained in living organisms. Large amounts are produced in fission: 10^5 Ci of strontium-90 and $1{\cdot}6\times 10^5$ Ci of caesium-137 are released for every megaton of fission energy, and it has been calculated that weapons tests up to 1962 have produced 9×10^6 Ci of strontium-90 and 14×10^6 Ci of caesium-137. Iodine-131 has a half-life of 8 d, but it is very concentrated by the thyroid. It and other short-lived isotopes, e.g. strontium-89 (51 d), are important only in local fall-out, for in the other two types of fall-out they will have decayed away largely or completely before the fall-out is deposited on the surface of the Earth. Most work on fall-out has been done on strontium, but it has provided much information which is relevant also to other nuclides.

It must be remembered that, in spite of the importance of the critical factors, others should not be entirely overlooked, for a change in conditions may make something critical where it was not before and, in addition, what may be quite unimportant in general terms may well be highly significant locally. An excellent example has been described recently (Preston and Jefferies, 1969). The Windscale reactor discharges its waste into the sea, and among the isotopes therein is ruthenium-106, the critical nuclide in this set of circumstances, which is concentrated by a seaweed *(Porphyra)* that is made into laver bread (pronounced 'lar-ver'). This critical food is eaten by people in the Swansea area of South Wales (the critical population group). The radiation from it is well below the MPC, but is nevertheless higher than in other foods, and is sufficient of a potential hazard to warrant regular sampling and monitoring of the seaweed and laver bread.

8.2 RADIO-ECOLOGY

8.2.1 General (Aberg and Hungate, 1966)

The object of radio-ecological studies is to determine the relationships between environmental contamination and harm to Man, in particular the relative importance of fall-out compared to natural radioactivity; and to find out the critical factors, which nuclides give most dose, and how much; and what remedial action, if any, is possible. But for a number of reasons this is not easy. The radiation doses concerned are difficult to measure precisely, for exposures and dose rates are low; techniques may have to be used right down to the limits of their sensitivity, and their reliability must be proved and checked at intervals. The distribution of radiation sources in the ecosystem is not homogeneous, so that sampling errors may lead to measurements that are not typical and are therefore misleading. There may be a long latent period between release of activity and its appearance in the diet, and radiation exposures both in the short term and the long term (the dose commitment: Section 6.5.3) must be considered (MRC, 1966). Large-scale surveys are necessary to determine the complete picture of the transfer of activity along food chains to Man, and all aspects should be covered, for changes in conditions and knowledge may mean that some factors become critical where they were not before, and vice versa. Despite ignorance on many questions, though, only a few nuclides are important generally, and even with these, the critical paths are few in number.

Food chains divide broadly into the *aquatic* and the *terrestrial*, though there is a little overlap in the consumption of aquatic animals and plants by terrestrial animals, notably by Man, the final link in the chain. The most significant difference, so far as contamination with radio-isotopes is concerned, is that fall-out can lodge on plants which Man eats, so that land food chains therefore present a much more direct potential route for ingestion of radioactivity into the body than aquatic food chains. Food chains in water are more complex than on land, the alternative name of food webs being perhaps more descriptive. This is especially so in the sea, though there are here short cuts in the form of filter-feeding organisms which can ingest particles, including those from fall-out. These organisms are both large and small, ranging from shellfish to whales, and some are of direct importance to Man. Nuclides become spread out in a food chain as they pass through it, since the organisms at higher trophic levels, e.g. Man, are less likely to be dependent on only one preceding level, and take part in the food chain at several points. These organisms also use comparatively more food for growth than for energy, so that materials required for energy will tend to be conserved longer than those needed for growth.

All food chains, especially those in water, concentrate activity as it passes along them, and this concentration, which can be by factors of up to some thousands, or even more, can result in hazardous levels in the succeeding members of the chain, of which the final one is Man. Thus, though radioactivity may be diluted quickly after its release, particularly in water, it may also be assimilated equally quickly by living organisms, and in consequence water containing so little activity that it is fit to drink may support life containing too much activity to be fit to eat. The build-up of activity in organisms is considerably easier to measure in aquatic than in terrestrial food chains, where it is expressed as the *accumulation factor* (*AF*), defined as the activity of an element in an organ or tissue compared to its activity in the same weight of the environmental water. *AF*s can be very high, although extremely high values must be interpreted with caution, for they may indicate that the organism contains more radioactive element than its own weight (Polikarpov, 1966)! *AF* values less than 1 indicate discrimination against the isotope by the organism or tissue.

8.2.2 Radioactivity and plants

Plants may be contaminated either *indirectly*, through absorption from the soil, or *directly*, by deposition of fall-out on the plant itself or by any route not involving the soil. Only soluble material can affect plants indirectly, whereas animals can ingest and possibly digest insoluble solids. Direct deposition contaminates the plant faster than indirect absorption; it is described as the *rate-dependent* component of plant contamination, since its extent and importance vary directly with the current rate of fall-out deposition. Indirect contamination depends on the total amount of radioactivity in the soil, and is therefore described as the *cumulative-dependent* component; the soil acts as a reservoir able to contaminate the plant long after the actual deposition of fall-out has ceased. Mathematical expressions have been derived to evaluate the relative contributions of rate and cumulative processes to the activity in a plant, and to make predictions as to how they will vary with time after a deposition of activity. Direct contamination has received less attention than contamination via the soil, since the factors determining it are more diverse and variable; results are therefore less clear-cut and less rewarding. In addition, a detailed investigation of direct contamination often necessitates the development of new experimental techniques, and sometimes it is difficult to design experiments to give meaningful information, whereas studies on the soil are able to adapt existing methods without difficulty.

There are two phases in the contamination of a plant by direct deposition from the atmosphere, somewhat overlapping and merging into each other: first, when contamination is predominantly superficial, and secondly after the contamination has been absorbed into the plant tissues. With continuous deposition, as opposed to a single fall of activity, the overall effect of both phases together is observed. Once inside the plant, direct contamination may or may not be distinguishable from that picked up indirectly from the soil, depending on the behaviour of the element concerned, except in so far as a time factor is introduced from the one being rate dependent and the other cumulative dependent.

Direct deposition of airborne contamination on plants is influenced by the physical form of the fall-out. Much is brought down in solution by *rain*, and it is therefore found that drier areas receive less than wetter regions. Fall-out in rain is best studied by simulation, i.e. by spraying plants with a radioactive solution, rather than by protection, i.e. by growing plants under cover, which can affect the microclimate significantly and give up to a twofold variation in the uptake of radioactivity in a season.

Particulate fall-out is retained on plants in inverse proportion to its size; particles above about 5 μm fall out comparatively rapidly, but most (99%) drop off plant surfaces likewise; smaller ones are held by surface forces, and about 20% may be retained. Very fine particles and vapours deposit preferentially at the edges of a leaf, because of the aerodynamic pattern of air flow round a flat surface, and much iodine-131 deposits in this way, though it is not clear how important it is in comparison to particulate deposition. Iodine vapour may also adsorb onto dust in the air and it will then, of course, behave in the same way as solid particles.

The form of the plant can significantly increase deposition if it provides regions where particles can lodge; the best example is seen in the grasses, in which the junction of the leaves with the stem makes an excellent collecting funnel.

Plants may be contaminated directly at three sites: the leaves, the flowers, and the plant base and surface roots, described respectively as *foliar*, *floral*, and *plant-base contamination*. The relative importance of the three sites varies with the contaminating isotope, the type of plant (whether it is annual or perennial) the method of husbandry, and the climate. Floral contamination is particularly important with cereals and grain crops, since the grain that develops from the flower is eaten. Perennials and permanent pasture suffer contamination more through the plant base than through the leaves, since a mat of organic detritus gradually develops on the surface which catches and holds contamination like a sponge. There is not time for this to happen with annuals, and in any case several annual crops are cultivated between the rows to eliminate weeds, which effectively breaks up any plant-base mat that has managed to form, and consequently decreases absorption of isotopes; inter-row cultivation may be difficult or impossible with perennials. Foliar contamination of plants whose aerial parts are eaten is the critical path for short-lived isotopes, since their half-lives are so much shorter than the time taken for entry via flowers and soil that they have effectively decayed away completely before reaching the leaves.

By no means all the activity initially deposited on a plant stays there, and that which is not retained cannot contaminate the diet of an animal eating the plant. A realistic measure of *initial retention* is given by the amount held on the plant for 1 d after deposition; this is typically about $\frac{1}{4}$ of the total deposited. A significant proportion of a particulate deposit is shaken off soon after deposition, by wind or by the movements of nearby animals. Just as rain deposits fall-out on a plant, so it also leaches it off, irrespective of the chemical properties of the nuclides; only a small proportion of the activity in solution is rapidly absorbed into the plant, and the remainder is washed away, if it is not ingested first by an animal. It is not fully clear whether the loss is determined by the frequency or intensity of rainfall or by humidity, or by a combination of these factors. In addition, different parts of the plant may vary in their ability to retain contamination, and the deposit may be redistributed as time passes. The extent of retention is often affected by the density of herbage per unit ground area. In arid climates, dehiscence of the plant cuticle may remove a certain amount of surface contamination.

The rate at which a single fall of activity is lost is expressed as the *field loss factor*, or half-removal time, the time taken for contamination to be reduced by a factor of two from causes other than radioactive decay. The factor is an exponential term, and seems fairly constant for different nuclides at about 2 weeks (UK average). Growth of the plant may affect it: a factor of 13 d, on a basis of unit ground area, drops to 8·5 d when based on unit weight of dry matter.

Continuous deposition of activity leads to a steady state in which loss from the herbage is balanced by fresh deposition. After about 2–3 months, direct contamination of the plant by a single fall loses its importance to indirect contamination from the soil, and field losses from flowers and foliage are offset by increased absorption from plant base and/or soil, whose contribution is determined by how much activity was deposited in the period up to 2–3 months before.

8.2.3 Radioactivity in the soil

The ions in the soil may be in true solution in the soil water, or bound exchangeably on the surface of the soil colloid, the organic matter, and clay minerals. These form the labile ionic pool, and comprise mainly calcium (especially in cultivated soil, from liming), magnesium, sodium, potassium, chloride, nitrate, and sulphate. These two categories of ions are freely exchangeable and freely available to plants. A third category is those that are bound non-exchangeably in the soil colloid; they are not available to plants and, since the colloid is virtually insoluble, they can be released only by weathering of the soil.

The availability of ions to plants (thinking particularly of radioactive ions) depends on several factors, including *retention* and *movement* in the soil. The retention in the soil is the tightness of the ion binding. Ions are more readily available in organic and lateritic soils (tropical soils, low in organic matter and clay) than in clay soils, adsorption being stronger on clay minerals (Goldsmith, Bolch, and Gamble, 1969). The amount of radionuclide that can be fixed increases with the soil particles' specific area.

Lateral movement takes place by surface wash, whose extent is governed by rainfall, topography, and the nature of the soil. Downward movement is due to mechanical displacement, in which the humble earth-worm is an important contributor, and to diffusion. It is a slow process in undisturbed soil; cultivation speeds it up considerably by mixing the soil more or less evenly to the depth of cultivation. Movement is slightly faster in calcareous loams than in acid clays, since (as just mentioned) ions are more tightly held on clay minerals. The depth of the absorptive roots of the plant has an important effect on the amount of radioactivity picked up from the soil for, even though the activity may not move to the roots, the roots may grow towards the activity. If the activity is in the surfac layers, then increased plant growth (arising e.g. from greater soil fertility) may tend to decrease the concentration of radioactivity in the plant, if the roots thereby penetrate more deeply. Dry conditions also encourage deeper rooting and may have the same effect. On the other hand, if the roots penetrate into a zone that is more highly contaminated, then the amount of activity the plant picks up will naturally tend to increase. Variations in root uptake can give large variations in the activity of a crop, even in a single season.

The uptake of activity from the soil by plant roots can be made use of in experimental studies on the tracing of root penetration in soil; the soil is labelled with isotopes of increasing half-lives in layers at increasing depths, and the time when each appears in the plant indicates how long the roots have taken to reach that region of the soil (Cohen and Tadmor, 1966). The isotopes used must be immobile in the soil.

Absorption of radioactive ions by the plant is affected by the presence of other ions. They may compete with each other, so that less radioactive ions are taken in than if other ions were absent; or alternatively, the active ions may be displaced from their adsorption sites in the soil colloid, so that the plant takes up more active ions than in the absence of other ions. Hydrogen-ion concentration (i.e. soil acidity) has no great effect provided that the activities of metal ions are unaltered, and that they are not precipitated.

The plant itself shows a certain degree of discrimination, selecting against monovalent in favour of multivalent ions at the stage of entry into the root; however, the preference is reversed when the ions are actively transported into the conducting tissue in the root centre.

The radioactive content of a soil may be determined without difficulty, but this is not necessarily the same as the amount available to plants. In order to find the availability of an ion, or (more important) the quantity a plant actually takes up, it is necessary to grow a plant in the soil and then analyse the plant. Soil-extraction methods may be able to indicate how tightly a given ion is held in the soil, but cannot be used to estimate plant uptake until they have been proved to give the same result as whole-plant analysis. It may be concluded that generally only a few percent of the radioactivity in soils enters the plants grown in it.

8.2.4 Radioactivity in animals

There are several routes by which radiation sources can enter the animal body (Section 6.1.4). For many nuclides, the gastro-intestinal tract is the most important route of entry into terrestrial animals, though inhalation and skin absorption can be important under certain circumstances, and should not be ignored simply because they are minor routes and less work has been done on them. Skin penetration depends on the chemical form and solubility characteristics of the isotope. Work on the inhalation of isotopes is difficult to do and to correlate with field conditions, and the results from different laboratories do not always agree with each other (ICRP, 1966).

The study of the transfer of activity from plants to animals is difficult in that it is only possible to find out what animals eat '*in vitro*', i.e. under controlled conditions indoors. In the field, they will select what they like, and will not necessarily eat different plants in the proportions in which they appear in the pasture, so that sampling the herbage has little value. Even if the pasture was sown with a single species, there are almost certain to be some weeds present, and the animals may prefer a particular part of the plant. With a mixed sward, it becomes impossible to find out what they have selected.

The alimentary canal of the animal is permeable only to soluble materials, so that activity will be absorbed into the blood-stream and thence pass to the rest of the body only if it is already soluble or is made so by digestive enzymes. Insoluble and indigestible fall-out will not be absorbed into the blood-stream, but it will irradiate the gut tissues as it passes through before it is finally eliminated. The region of the gut in which absorption takes place can be determined by noting at what time after ingestion the isotope enters the blood-stream at the fastest rate, and comparing this with the known speed of passage of food through the gut. Some materials are absorbed by passive diffusion and others by active transport, with corresponding differences in rate and extent of absorption.

The composition of the diet can affect the uptake of activity into the body, the best example being the inverse relation of strontium and calcium (Section 8.3.1); the concentration of stable element with respect to the radioactive can also affect the uptake, but its influence is often difficult to evaluate. Absorption is generally more efficient in young animals than in older ones. Though the quantities of nuclides absorbed do not show much species differences (given the same amount in the diet), the rates of absorption and excretion may vary greatly. The activity taken up with the food will obviously increase with its isotope content, and this may be influenced by habitat and feeding methods, e.g. birds gathering their food from trees are likely to pick up less radioactivity than those grubbing it up from the ground and picking up contaminated soil with it.

How much isotope is eventually deposited in any particular part of an animal (the edible tissues being of most concern) is determined by the net result of the several physiological processes competing for the isotope; some nuclides are selectively concentrated in their critical organs (Section 6.1.4), and distribution within one organ may not be uniform, even though the organ may be biochemically homogeneous otherwise. Distribution and retention are affected by many factors, which are not yet fully defined; they include age, pregnancy, season, diet, genetics, diseases, and possibly sex. Ruminants show differences from non-ruminants, and turnover may be faster in smaller animals.

Once within the body, the behaviour and distribution of an isotope is usually the same, regardless of its entry route. It may exist as the free ion or in a complex of some form, and it may be absorbed by an organ and reappear subsequently in a changed chemical form. Protein–metal complexes are thought to be important in the storage, transport, and excretion of isotopes.

Under conditions of steady intake, the concentration of isotope in a tissue will at first increase and then level off to a plateau. When ingestion of activity ceases, it is gradually eliminated, provided that it has not become permanently fixed, as may happen in the growing animal. Retention in the body is described by the sum of two exponential functions, the shorter relating to the loss of unassimilated nuclide from the gut, and the longer one describing the excretion of assimilated nuclide from the tissues. This gradual loss of endogenous nuclide may be a complication in studies of the absorption and retention of a single experimentally administered dose for, if some of the same isotope is already present in the animal, an erroneously low result is likely.

Except at very high concentrations, far and away above the maxima permissible in food for human consumption, the presence of radio-isotope in an animal or plant has not been shown to have any adverse effects on its normal functioning (e.g. Sparrow and Puglielli, 1969). Although the presence of isotope in food may make it undesirable or unfit for human consumption, its nutritional qualities as a foodstuff are not otherwise affected, i.e. it is just as nutritious as inactive food.

8.2.5 Radioactivity in water

The aquatic environment is very different from land, largely due to the physical properties of water, and the differences naturally influence the distribution and spread of activity through it. Water ecosystems embrace a wide area of the Earth's surface and an equally wide range of organisms, and relatively little work has been done on them; knowledge is largely empirical, and it is difficult to generalise, but there are a few fundamental concepts. Watersheds can be important for, though small in area, they handle a large amount of water.

As already noted, food chains in water are more complex than on land, and there is an additional source of radioactivity as compared with terrestrial food chains, arising from the disposal of radioactive wastes into rivers and the sea, which is most important near the coast. Some activity may be induced in reactor cooling waters by neutron capture, but the hazard it presents, if any, is very localised. The wastes (and induced activity) in one body of water are not necessarily the same as those in others, and each must be studied individually. In assessing the hazard to Man from activity in water, the major critical factor is waste discharge, if present, and aquatic food chain studies are therefore concerned with this as well as fall-out, although in the sea a natural isotope, potassium-40, is quantitatively predominant.

Polikarpov's monograph (1966) on the radio-ecology of aquatic organisms is a useful survey. It concludes that the concentration of long-lived fission products in the marine environment is continually rising, and artificial radioactivity is now a permanent feature of the sea; it states that several countries, including the United Kingdom, France, and the United States, but not the Soviet Union, are dumping increasing amounts of radioactive wastes in the sea. Damage to nuclear powered ships and to reactors on the coast could have extremely significant consequences, and the numbers of such ships and reactors is increasing. However, it must be pointed out that British and American waste discharge is authorised only for certain low-activity material after careful consideration, and that high-activity material, which makes up some 99% of the waste, is permanently stored.

There are three basic research trends in marine radio-ecology: first, analysis of the radio-ecological situation, i.e. of the content of isotopes in the environment and in the important marine organisms; secondly, prediction of the accumulation of nuclides and their release by plentiful organisms; and thirdly, prediction of the threat to biological productivity and marine life from contamination of the sea. These trends cover both the effect of organisms on the environment and the effect of the environment on the organisms.

Radioactivity is quickly diluted and translocated in water, more so than on land, and especially in flowing water. Translocation takes place through bulk water movement (currents), by sedimentation of particulate matter, and in association with living organisms. In the sea, the spread of activity is more complex than in fresh water, simply because there is so much more of it; distribution of activity is not uniform for, though all the seas and oceans are interconnected, some areas mix little or not at all with the rest and, within one ocean, uniform mixing is achieved only after a long time, if ever. Horizontal movement of the sea is chiefly under the influence of wind, and is clockwise in the northern hemisphere and anticlockwise in the southern. Vertical mixing is extensive only in the surface layers down to about 100 m depth; it is assisted by currents in coastal regions, and is hindered by density differences that arise from temperature and salinity gradients in the water. At greater depths, there is a slow down-welling of water at the Poles and a corresponding slow rise near the Equator, with residence times estimated at up to about 600 years in the Atlantic and about 1000 years in the Pacific, though there is evidence for a much more rapid vertical exchange between the bottom and surface waters in the Pacific and the Black Sea, of about an order of magnitude faster. Currents may also bring radioactive substances to the surface in a far shorter period. Complete mixing of the surface water may take between 6 weeks and 1 year, though most particulate fall-out settles out within a few days. Sedimentation may ultimately remove, depending on the element, from 10 to 60% of the activity in water; it will then be available to the bottom-living biota. However, strontium is not adsorbed by marine sediments, especially at acidic pH, and it may remain in sea-water for a very long period, until complete decay, and be available to organisms.

Nuclides become associated with living organisms by adsorption on to or absorption through surfaces, and by ingestion with food. Contamination increases with the specific activity of the isotope and with the physiological demand for the element. Aquatic plants and plankton may take up activity rapidly, within a few hours to a few days and, having once accumulated it, some organisms may lose it only very slowly. Little seems to be known of how the uptake of an element by aquatic organisms is affected by its physical and chemical state, by the presence of other substances in the water, or other factors, such as the significance of detritus formation in the migration of radioactive substances. For example, dissolved organic compounds may chelate metallic ions and decrease their availability. Uptake of activity by larger animals may be the result of its uptake by bacteria associated with the animal (Ahearn, 1968). An indirect effect is that an element in short supply (often phosphorus or nitrogen) may be the limiting factor in the growth of the biomass and therefore in the extent of contamination. A further complication is the variable composition of fresh water.

The fact that some organisms concentrate radioactivity from water to a great extent means that they can act as *indicator organisms* to reveal the presence of low levels of contamination in water at or below the limits of detection if the water were monitored directly. It has also been suggested that such organisms could be used to decontaminate large volumes of water intended for human use, or liquid waste, although on the other hand the presence of algae in settling ponds for the disposal of radioactive liquids has been troublesome (Echo and Hawkins, 1966). Information on the concentration factors of a number of nuclides by marine organisms is still lacking.

In spite of inadequate study of many aspects and many unanswered questions, the main features of aquatic radio-ecology are fairly well outlined in general terms. Food chains are apparently of secondary importance in the concentration of radioactive matter by marine organisms, in contrast to the situation on land, and concentration is by direct uptake from the water. Although plankton account for only a small part of the activity entering sea-water, they can nevertheless function like a pump in transferring nuclides to deeper levels of the sea or to the bottom.

More research is needed on the maximum permissible concentrations of activity for marine organisms, especially in the hyponeuston, the large amount of life in the uppermost two inches of sea water. It is the largest ecosystem in the world, and it is probably the most vulnerable to radioactive fall-out. Damage to it may seriously disrupt reproduction of commercially important fish, many of which begin life as free unprotected ova, floating near the surface and readily capable of adsorbing quantities of many nuclides. The radiation hazard to sea life is made worse by the fact that in the past it has received less natural background than terrestrial organisms, so that the evolutionary process in the sea may be much more easily altered by excess radiation, though little has been published on the radiosensitivity of marine plants and animals. Polikarpov (1966) concludes from data on radiation effects on anchovy eggs and fry that further radioactive contamination of the oceans is inadmissible. His results have not been confirmed by other workers, though, and they are somewhat remarkable in indicating that, while significant abnormality production begins at dose rates only a few per cent higher than those from the potassium-40 natural background, the abnormality production rate only increases by a factor of three when the dose rate is increased by seven orders of magnitude. This implies that a biological system can exhibit an extraordinary radiosensitivity at very low dose rates, and a very high radioresistance at high dose rates, and in consequence most experts in the field are very sceptical about Polikarpov's conclusions (Morley, 1970).

8.3 INDIVIDUAL ISOTOPES

8.3.1 Strontium

Strontium and calcium are metabolically interdependent, being adjacent members of the alkaline-earth group of elements, and consequently it has been found that a truer picture results if concentrations of strontium are expressed in terms of its ratio to calcium, the Sr/Ca *ratio*, rather than simply as absolute quantities of strontium. The unit used is pCi of Sr per g of Ca. However, though the behaviour of the two elements is similar, it is not identical, for several physiological processes discriminate against strontium in favour of calcium; the strontium ion is bigger than calcium, and therefore does not fit the calcium transport mechanism correctly. The *discrimination factor* (*DF*) for a single physiological step expresses the decrease in the Sr/Ca ratio that takes place at that step. Where there are several steps in an overall physiological process, as for example in the transfer of strontium from herbage to milk, the overall discrimination in the conversion of precursor to product is the *observed ratio* (*OR*), the Sr/Ca ratio in the product as compared with that in the precursor. The physiological discrimination against strontium has the important consequence that strontium is concentrated much less than it would be otherwise in its passage along the food chains from soil to Man.

The relative importance of strontium-89 and strontium-90 varies considerably with time, because of their different half-lives, 50 days and 27 years respectively, through which strontium-90 gradually becomes more significant. The ratio of strontium-89 to strontium-90 in fall-out is about 10 : 1, but is more than halved in the diet because of the time taken to pass through the food chains. Strontium-90 gives much longer-lasting effects; the dose from it is about twice that from an equal activity of strontium-89 initially (because of its harder radiation), about 10 times after 1 year, and 150 times after 20 years; the dose commitment is more than 400 times greater.

Strontium in the soil

The availability of strontium in the soil is about 10 times that of caesium; it is essentially all in the labile ionic pool, albeit on surfaces rather than in true solution, and little or none is irreversibly fixed. It can therefore enter plants more than caesium, but only about 5% of the quantity in the soil in fact does so because of the discrimination against it in favour of calcium. The relationship between the two elements means that plants tend to absorb more calcium than strontium, except in calcium-low soils containing less than 10 mEq exchangeable Ca/100 g soil, from which strontium absorption is increased appreciably, giving an inverse relationship between soil calcium and the Sr/Ca ratio in the plant. Liming of calcium deficient soils will therefore decrease strontium absorption from them. The presence of stable strontium does not affect the uptake of strontium-90, since its effect is overwhelmed by the much greater amount of calcium present. Since the behaviour of strontium is governed by that of calcium, it is important to determine the relative availabilities of the two elements in a soil, and not the availability of strontium in isolation. The effect of the soil type is less important than for caesium (Section 8.3.2), but less is absorbed by plants growing in clay soils, since strontium is held more than calcium on clay minerals. Some strontium may also be retained in humus horizons, and some may become fixed and non-available after a period of time in the soil. Movement of strontium is more rapid than that of caesium: it may penetrate to a depth of 2 ft within 3 years. Soil moisture does not seem to affect the uptake of strontium by plants, though it may be that soil factors are less important than plant factors in modifying strontium uptake from soil, be it wet or dry. High-temperature treatment of strontium-contaminated soil (800–1000 °C) reduces uptake, especially from clays, and this has been suggested as a remedial method for use on such soils, though the question of its practicability on a large scale is open to debate.

Strontium and plants

There is some discrimination against strontium as it penetrates into the xylem duct in the root, but the *OR* of plant : soil is approximately 1, although it does vary in different parts of the plant.

The initial retention of strontium on pasture foliage has been estimated at 15–30% (average 22%) of the total deposit per unit area. The field loss factor is 50% in about 2 weeks, expressed per unit ground area, or in 9 days, expressed per unit dry weight of herbage. Little or no strontium enters the plant from direct contamination of the leaves; it is relatively immobile in the plant and downward movement is negligible away from the site of contamination. Thus crops whose edible parts are protected by outer leaves (e.g. lettuces and cabbages) are easily freed of direct strontium contamination by removing the outer leaves and with them the contamination; the crop may then be eaten without harm. Likewise, root crops are almost unaffected by direct deposition. Direct foliar contamination with strontium is therefore relatively unimportant by comparison with the plant base, though it seems of greater relative importance in pastures of low productivity. Strontium reaches the plant base both by direct deposition and by leaching from the foliage. In permanent pastures especially, it may be retained for a considerable period in the mat of prostrate stems, organic detritus, and roots at the soil surface; and absorption from the plant base reflects the fall-out deposition of about 2 months previously. Floral contamination is most important in cereals at times of appreciable fall-out, resulting in a higher Sr/Ca ratio in the grain than in other tissues.

Strontium in animals

Strontium absorption from the animal gut is passive, whereas calcium is taken up by active transport, so that only some 20–50% of the strontium intake passes into the body. As dietary calcium increases, more strontium is excreted in the urine (Fujita, Iwamoto, and Kondo, 1969); age may also influence strontium retention in the body (Anderson and Comar, 1968). Strontium deposits chiefly in milk and in the bones. The alkaline earths in bone are to some extent exchangeable, with a turnover of 100% in the first year of life, and about 1% per annum (thigh bone) to 8% (vertebrae) in adults. The bones thus act as a reservoir from which strontium may be slowly released and recycled, appearing in milk and elsewhere, over a long period of time. Its biological half-life is about 17 years. The *OR* for milk : diet in cows is about 0·1 and the *OR* for body : diet in man is about 0·25. Addition of alginate to milk reduces strontium absorption from it, presumably through the formation of a non-diffusible chelated complex. Calcium metabolism is highly important in pregnancy and lactation; a slightly negative calcium balance in late pregnancy will not release a bone-seeking element (alkaline earth) trapped in the bones before pregnancy, but, in lactation, when calcium is mobilised even more, it will do so. Alkaline earths ingested in pregnancy do not show any preference for selective deposition in the foetus and placenta or in the mother (Sternberg, Legave, and Marcil, 1969). Infants and young children are the *critical population* group for strontium contamination and measurements of their strontium concentrations are constantly being made, e.g. Levi *et al.*, 1969. Ebel and Comar (1969) have suggested the use of Sr/Ca ratios in hair for monitoring.

Strontium in water

Strontium and other bone-seeking elements, the alkaline earths, enter *fish* through membranes (the gills) rather than via the gut. Studies with the herbivorous fish *Tilapia* show accumulation of strontium in its organs in the order: skeleton $>$ skin $>$ gill $>$ $>$ muscle $>$ viscera. Accumulation factors are inversely related to the calcium content of the water, and expression of data as *OR*s is more satisfactory, underlining the importance of considering strontium in association with calcium and not in isolation; for a meaningful comparison of *AF* values with each other, the calcium concentration of the water should also be known. There is little discrimination against strontium by fish, the *OR* for body : water being about 0·5, but much accumulation, *AF*s are in the range 500–10^5 depending on the calcium in the water.

In the *sea,* strontium is the most hazardous fission product, though in general human exposure to it from sea produce is relatively small compared with terrestrial sources. Concentrations in sea-water in 1959 were reported as 0·04 pCi/l in the North Atlantic, and half this in the South Atlantic, compared with 0·01 pCi/l in fresh water. In 1959–60 the Windscale and Calder Hall installations were discharging it at the rate of about 1500 Ci/year. Concentrations have risen since, and have reached 1 to 100 pCi/l and above in the Pacific Ocean and the Irish Sea; Polikarpov (1966) suggests that they should not be allowed to rise any higher (but *see* end of Section 8.2.5). The specific activity of strontium is lower in the sea than on land, since sea-water contains about 8 mg/l of the stable element; and it is considerably more uniformly dispersed. Its effects are lessened by the presence of calcium and by discrimination, but nevertheless accumulation factors can be high, e.g. filamentous algae 500 000; phytoplankton 75 000; insect larvae 100 000; fish 30 000–70 000 (Kalnina, 1969). *Acantharia*, brown algae, and the shells and bones of marine animals are suggested biotic indicators.

8.3.2 Caesium

As with strontium and calcium, caesium and potassium are metabolically interrelated and, though the relationship is not as close as that of the former pair, it is close enough to justify the use of Cs/K *ratios*, expressed as pCi of Cs-137 per g of K.

Caesium in the soil

When considering caesium in the soil, three broad grades of soil may be distinguished. Mineral soils are found in the majority of agricultural areas, and are characterised by a low or moderate organic matter content with much clay mineral. Caesium is strongly and rapidly fixed in the clay mineral lattice, most strongly with micaceous minerals; less than 1% of caesium in the soil is in solution, and its availability to plants is 10% that of strontium. The tightness of its fixation is reflected in its very slow movement down an undisturbed soil profile, typically about 1 in in 3 years, though somewhat faster in a calcareous loam than in an acid clay soil. Fixation is progressive over the first 2 years after deposition, and its availability is then down to not more than 2·5% of that of strontium. Its annual contribution to dietary contamination is therefore small, though it may be appreciable over a long period.

There is far less information about availability of caesium in the other two types of soil. Lateritic soils occur in the tropics; they contain very little clay mineral and are also low in organic matter, and caesium is more readily available than in mineral soils. Caesium also enters plants readily from organic soils; these too contain little clay mineral, and the high-organic-matter content can reduce the capacity of the clay to retain caesium, though the mechanism is little understood. This is the only explanation as yet for the high amounts of caesium in Arctic plants, which grow on soil relatively high in organic matter. Under permanent pasture, a peaty organic layer may accumulate in the surface horizons, and this might be able to bring about the freeing of some of the caesium from its binding by clay mineral and account for the entry of caesium into pasture plants that is sometimes observed a year or more after its deposition. This making available of caesium can cut the other way, though, for if a mineral layer underlies the organic, any of the now-mobile caesium entering it will promptly be re-fixed.

The presence of other ions decreases caesium absorption from the soil: potassium very markedly, sodium, calcium, and magnesium much less, and ammonium less still. The effects are best seen by addition of the ions to soils low in them. Thus in mineral soils, especially where potassium is present, the absorption of caesium is negligible in comparison to other ions. Total chemical caesium in the soil is normally very low, and addition of the stable element enhances the absorption of the radioactive ion, since it displaces it from its fixation sites. Under dry soil conditions, caesium in the plant is increased considerably, and potassium but slightly; both accumulate more in stems than in leaves.

Caesium and plants

The cumulative deposit of caesium in the soil makes little contribution to entry into food chains, most food coming from cultivated mineral soils, although caesium can sometimes enter plants relatively freely. Rate-dependent contamination is much more important, and it should be noted that with caesium this is not synonymous with 'direct contamination', for some caesium may reach the plant after a brief sojourn in the surface layer of the soil. There are fewer detailed studies than with strontium of the two phases of rate-dependent plant contamination, e.g. on the contamination of permanent pastures, other than fall-out surveys which represent the combined effect of both phases. However, the initial retention of caesium on plants and the field loss factors from them seem to be similar to those for strontium, so that there is some common basis for comparing their uptakes by plants and animals.

Plant base absorption is the major route by which caesium passes into plants, though it may operate less efficiently than for strontium. In the localised situation at high latitudes, entry of caesium into plants is greater than in the more-familiar temperate regions, and takes a different time course. In these cool and Arctic regions, the soils, especially the surface layers, are much more organic, and availability of caesium after it has entered the soil may be more important than absorption via the plant base. The flora is based on lichens, and much more work is needed for a full interpretation of their caesium uptake.

In contrast to strontium, caesium is *mobile* and readily translocated throughout the plant, so that direct deposition on the aerial parts leads to widespread contamination everywhere else; subterranean storage tissues may contain tens or hundreds of times more caesium than strontium. This mobility of the alkali metals can be put to good use: after administration of isotope to the shoots or leaves, the plant becomes uniformly labelled, and the activity in soil cores taken around the plant gives a measure of how far and how deep the roots penetrate, much more easily than the conventional technique of washing out the plant from the soil.

Caesium in animals

Uptake of caesium from the animal gut is extensive, 70–100% of the amount taken in, and it is rapidly distributed over the whole body. Its retention varies with age and sex (Boni, 1969); its biological half-life is about 110 d in Man and about 30 d in cattle. Ruminants retain somewhat less than non-ruminants. In cows, the *OR* for milk : diet is 0·1, and it reaches milk some 10 times more readily than strontium. Vandenhoek *et al.* (1969) found a highly significant seasonal effect on the caesium in milk, there being up to three times more in late summer than in early summer, in contrast to the relatively constant strontium content. As might be expected, the caesium in the milk was more variable on free grazing than in more closely controlled feeding systems.

Caesium in water

Plankton remove caesium from water fairly quickly; significant amounts are taken up within 2 h, 95% of a single dose is removed within 50 h (2 d), and 99% within 5 d. Estuarine organisms show a seasonal variation in caesium uptake (e.g. Wolfe, 1967), perhaps due to corresponding variations in radioactive deposition and/or in salinity of the river water. Accumulation factors may be quite high, e.g. bulrushes 90; duckweed 500; tadpoles 1000; carp 3000; plankton 1000–25 000.

Caesium is rapidly diluted in the sea, but it has a relatively high specific activity since there is only a little of the stable element present; however, the relatively large amount of potassium serves to lessen its effects, and it does not appear to be a fission product of major concern in the oceans. Brown and red algae and the soft tissues of marine animals are suggested as biotic indicator organisms for caesium.

8.3.3 Iodine

The main property of iodine-131 determining its characteristic features as a dietary contaminant is its short half-life, leading to great variability and rapid changes in fall-out and contamination levels that are typical of short-lived isotopes. In addition, the amount present in the diet will be greatly affected by the seasonal nature of agricultural production, and the delay between production and consumption of many foods will effectively reduce it.

Iodine is unimportant in stratospheric fall-out, which comes down to earth slowly, and it is a constituent of local and (somewhat less) of tropospheric fall-out only. The fall-out pattern is therefore affected by weather conditions, which again impose variations on the amounts deposited. Likewise, iodine has no significance in the soil and in water, and *direct foliar contamination* of terrestrial plants, over a relatively short time period, is the major critical path, if not the only route it takes, through food chains; it decays away so quickly in comparison with the time taken in the other routes that effectively all of it gets lost on the way.

Iodine is readily absorbed (70–100%) from the animal gut; most of it finishes up in the thyroid gland, with some 5–10% getting into the milk of a lactating animal. Equations have been developed to estimate the amount of radio-iodine consumed by animals after a single contaminating event, and its relation to that taken in by Man in milk. The collection and analysis of bovine thyroids, as compared to milk sampling, is very sensitive for detecting environmental intrusions of iodine-131 in space and time. Diet affects iodine uptake from the gut; the stable iodine content, and also renal iodine loss, seem important factors.

It is possible that some iodine fall-out may enter the body by inhalation and percutaneously, as well as in the food, since it is often in the form of very fine particles or vapour. Using the miniature pig, which is comparable to Man, as the experimental animal, it was found that only 10% of the amount inhaled as vapour entered the body through the unbroken skin, and much less if the isotope was particulate. With sheep, the amount on the wool was barely detectable (Morgan *et al.*, 1968).

8.3.4 Other nuclides

Fission products and induced activity

Other than those in the preceding three sections, fission products are almost always comparatively unimportant, because of either their very low yield, or short half-life, or immobility. They may be taken in with food as adsorbed material or solid deposit on plants or inside smaller animals eaten whole, but they do not concentrate, with minor exceptions, and they are generally limited to low trophic levels (Gueguenlat, Bovard, and Ancellin, 1969). There may be surprises, however, e.g. plants grown on the ejecta from a thermonuclear excavation contained radio-tungsten as the dominant nuclide, and strontium, caesium, and iodine were relatively minor.

A little induced activity results from capture by existing stable elements of the neutrons released in fission. It may be important locally, but no more, in water, and it can be ignored on land. The isotopes formed by (n, γ) reactions from the corresponding stable isotopes of the same element include zinc-65, cobalt-60, iron-55, and iron-59. (The commonest stable iron nuclide, Fe-56, generates another stable nuclide, Fe-57; its abundance is 90%, whereas that of Fe-54 is 6% and that of Fe-58 is 0·3%.) Manganese-56 and sodium-24 are induced by the same reaction but, though they are biologically important elements, their half-lives are short, $2\frac{1}{2}$ h and 15 h respectively. Manganese-54 may be formed by an (n, p) reaction from iron-54.

These elements may be concentrated greatly to high specific activities by biota, including some economically important fish and invertebrates. The concentration process for iron-55 by salmon and tuna fish gives 20–30 times as much as in caribou and 100 times as much as in beef, and people in northern countries (e.g. Scandinavia and Japan) who eat much ocean fish are liable to have relatively high body burdens of iron-55. Burdens of up to 1·5 μCi have been measured, but are not considered to approach maximal levels or to present a very great hazard (Palmer, Beasley, and Folsom, 1966). The freshwater clam has been found to be a good biological indicator for radio-manganese, pelagic fish eggs and algae for rare earths, algae for Zr-95 and Nb-95, and algae and crustaceans for Ru-106.

Nuclear reactors may give rise to some induced activity in the cooling waters discharged from them, e.g. Ru-106 (Section 8.1.2), though some reactors recirculate their water and do not release it. The dominant nuclide produced is phosphorus-32. Plankton are highly important in uptake from the water and subsequent passage to other biota; they remove phosphorus-32 quickly, much within 15 h. Accumulation factors may be high, e.g. zoöplankton 40 000; fish (*Fundulus*) 13 000; sponge 4500; sphagnum moss 400; *Daphnia* 100. There is evidence that some fish in rivers downstream from certain reactors are declining in numbers, due to shortened life span and retarded development arising from their uptake of radio-isotopes.

Natural activity

Potassium-40 is the most significant, but it is under homeostatic control in living organisms and there is therefore little benefit from studying it in food chains, though the quantity in infants may have some relationship to nourishment (Garrow, 1965). Ashby *et al.* (1967) observed a large increase in environmental radiation at the start of spring growth, from about 1 mR/week in winter to 3·5 mR/week in summer, and concluded it was due to potassium-40 in the developing foliage.

Tritium and carbon-14 are also widespread, both in nature and in the body, and they are in equilibrium throughout except where they are fixed, e.g. in the skeletal structures of the growing animal, and there is some point in studying them here. Tritiated water applied to the surface of the soil in a tropical rain forest (significantly a clay soil) was still present in the top 180 mm up to 7 months after application, so that, even in such a high-rainfall environment, plant roots may be exposed for a considerable time after the release of radio-isotope. The washing-out of the tritium from the soil showed a peak, followed by a long tail, due to trapping of water in isolated compartments in the clay. On a more freely permeable and less clayey soil, wash-out would doubtless be much more complete and quicker (Kline and Jordan, 1968). Tritiated water is freely absorbed through the skin, and even as vapour as much may enter the body this way as through the lungs. An isotope effect has been noted in algae for deuterium and tritium; their utilisation was half that of hydrogen (as water), so that in effect they were discriminated against. Harkness and Walton (1969) discuss carbon-14 in the biosphere and humans.

Although uranium–thorium series isotopes are very localised in their distribution, some study of these elements (and of the artificial transuranics, e.g. Turner and Taylor, 1968), is perhaps desirable, in view of their extreme toxicity and long half-lives. For example, sufficient uranium may be absorbed through skin contact to present signs of poisoning, and uranium processing and milling plants may give rise to appreciable amounts of radium-226 in the nearby environment. Assessment of the total alpha activity is the easiest method, but it is desirable to identify the individual isotopes, since their metabolic behaviour is different, and this may be done by a combination of chemical separation and spectrometry of the radiation. These isotopes do not enter plants, except for a small amount of radium, which seems to be discriminated against in favour of calcium more than strontium. Some lead-210 and polonium-210 may deposit on plant surfaces from decay of radon in the atmosphere, but are far less important than strontium, iodine, and caesium. The degree of uptake of ingested polonium-210 may depend very strongly on its chemical form; it may be less well absorbed in inorganic form (e.g. as fall-out on vegetation) than if organically bound, as in meat.

An interesting observation on cattle is that polonium-210, radium-226, and thorium-228 are concentrated 100 times more in the choroid and iris of the eye than in the non-pigmented eye tissues, and the resultant radiation dose in these tissues may be about 5 rem/year. The epithelial cells of the choroid are extremely susceptible to neoplastic change and, since squamous cell carcinoma of the eye is the commonest neoplastic disease in cattle, the question is raised, is this an instance of a form of mammalian cancer caused by natural radiation?

Natural radiation also affects the animal population in high uranium–thorium areas, and three population groups can be distinguished according to whether they have high, moderate, or low contact with radiation. The first group comprises digging and aquatic animals permanently inhabiting a high radiation zone; they may be exposed to 5–8 mR/h. The second group live in trees and descend to the ground to feed, where they may eat highly contaminated food; the third group are permanently arboreal types such as squirrels and birds.

8.4 CONTAMINATION OF THE DIET

8.4.1 The critical foods

The possible routes for the entry of radionuclides into Man's diet may be represented in the form of a simple diagram (Figure 8.1), and from what has gone before it is not difficult to predict what will

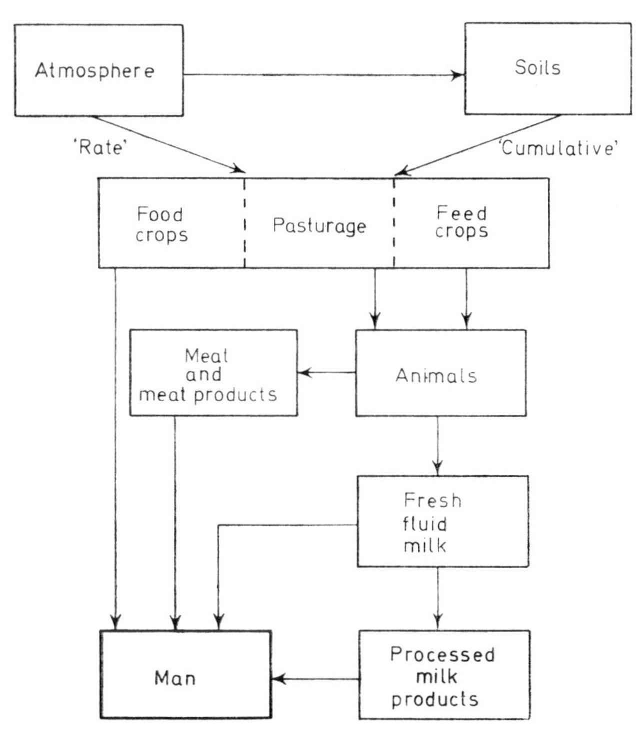

Figure 8.1. Critical terrestrial pathways of dietary contamination. (From Radioecological Concentration Processes, *by B. Åberg and F. P. Hungate, 1967). Courtesy Pergamon)*

be the critical paths and critical foods for the three nuclides of most concern.

Iodine-131. This shows the simplest pattern, with only one critical path; one food, fresh milk, is of prime importance, though some iodine may be taken in through the consumption of fresh vegetables, and some also appears in eggs. Iodine is significant only in the short term, and disappears if the food is stored for any appreciable length of time, in contrast to the other two nuclides; the amounts in the diet also show considerable, and rapid, variations.

Caesium-137. Here the major dietary sources are (in order) meat, milk, root vegetables, grain products (flour), tea, and leaf vegetables. Caesium-137 enters diet mainly by direct plant contamination, and the amounts in milk and the human body are directly related to the rate of fall-out (Iinuma, 1969). Entry via the soil is unimportant, though the activity fixed in the soil may be passed slowly into food over a long period. Caesium is the most important fall-out isotope in Arctic ecology, from the viewpoints of rates, routes, and amounts of isotope cycled. The critical food chain is lichen–reindeer (caribou)–Man. Reindeer meat may contain 10–100 times more caesium than that from animals in temperate zones.

Strontium-90. This enters into the diet in ways determined both by the current fall-out rate and cumulative deposition in the soil. Strontium-90 can appear in practically all foods, with milk as the major source, especially of direct contamination. Relative to the average total diet taken as 1, the Sr/Ca ratios in different foodstuffs are as follows (stable strontium and radioactive strontium do not give the same results):

	Milk	*Grain*	*Root Vegetables*	*Leafy Vegetables and Fruit*
Stable Sr/Ca	0·2	1·4	4·5	3·7
Active Sr/Ca	0·75	1·6	1·9	2·0
Calcium %	0·13	0·04–0·10	0·02–0·06	0·15

The typical United Kingdom diet has dairy produce as the main source of calcium, but differs from other 'Western'-type diets in that little whole-meal grain is eaten, it being replaced by white bread 'fortified' with mineral calcium; this lowers its Sr/Ca ratio by more than 25%. At times of high fall-out, dairy produce contributes $\frac{1}{2}$–$\frac{3}{4}$ of the strontium in the diet, reflecting direct contamination, and the rest comes from root and leaf vegetables and fruit; other things contribute very little. When fall-out ceases, vegetables contribute relatively more strontium, reflecting cumulative deposition in the soil. There is very little information about other types of diet, in other countries overseas, in which half or less of the calcium is provided by milk, with vegetables being of equal or major importance. The seasonal pattern of fall-out has a large effect on the relative extent of contamination of crops grown at different times of the year.

Milk. This is a source of all three isotopes, and the major one for iodine and strontium. The consumption of radio-isotopes by humans through milk and dairy products depends on the intake by the cow, loss from the pasture, concentration in the milk, quantity of milk (or its equivalent) consumed, and (for iodine) radioactive decay in the time since contamination took place. The milk yield is important in determining the concentrations of isotopes in it, which are for this reason best expressed in terms of the percentage of the ingested dose appearing per l of milk. Results for strontium are more meaningful in terms of the Sr/Ca ratios in diet and in the milk. The time scale for the contamination of milk is conveniently summarised in tabular form.

	Iodine	*Strontium*	*Caesium*
After a single dose:			
Isotope appears within:	30 min	—	10 min (in goats)
Maximum concentration appears in:	6–12 h	2 d	1–2 d
Its value (% of dose/l) is:	0·5%	0·02%	0·3%
Concentration then declines, with half-life of:	16 h initially, then more slowly	≈ 3 d	1 d initially, then 4 d, then at rate lost from muscle
If intake is continuous, at the same rate each day:			
Maximum concentration appears in:	4 d	6–7 d	7 d
Its value (% of daily dose/l) is:	0·3–0·5%	0·09%	More than 1 (cows) Up to 30 (goats)

A total of 6–12% of the daily intake of caesium by cows appears in the milk and 6–25% in goats, but because the milk yield of goats is much lower, a correspondingly higher proportion of the dose appears per l of milk. If cows are removed from pasture contaminated by a single event to uncontaminated feed when the peak concentration appears in milk, the milk concentration is:

	Iodine	*Strontium*	*Caesium*
Reduced 10 times in:	3 d	8 d	10 d
Reduced 100 times in:	10 d	21 d	49 d

If cows remain on pasture contaminated by a single event, the milk concentration falls off exponentially, and is:

	Iodine	*Strontium*	*Caesium*
Reduced 10 times in:	23 d	76 d	64 d
Reduced 100 times in:	43 d	154 d	133 d

Although milk is the main source of iodine in the diet, the concentrating capacity of the cow's udder is variable and small; considered as a machine for the transfer of iodine from diet to milk, the dairy cow is the least efficient of the domestic ruminants. Iodine in milk is affected seasonally, being highest in spring and summer, no doubt correlating with the change from stored winter food to fresh fodder. Increased thyroid activity can stimulate lactation.

Meat. Caesium is taken up slowly into muscle, the rate being dependent on the activity of the muscle. Equilibrium is attained about 10 d after the start of ingestion in the goat, and after at least 30 d in the cow. Meat may contain on average 3–5 times as much caesium as milk. Very little iodine appears in meat, about 0·001% of the daily (continuous) intake per g in the cow, about 0·003% per g in the sheep, and 0·0004% per g in the hen.

Eggs. Here again the picture with iodine-131 and caesium may be tabulated.

	Iodine	*Caesium*
After a single dose:		
Maximum concentration appears in:	6 d	1 d
Its value (% of dose/egg) is:	5%	1%
If ingestion is continuous, at the same rate each day:		
Maximum concentration appears in:	6 d	6–7 d
Its value (% daily dose/day) is:	8%	2·3–3·3%

The white contains 2–3 times as much caesium as the yolk, with 1–2% only in the shell; the yolk contains 20–50 times as much iodine as the white or shell. The period of peak concentration of iodine by the developing oöcyte is about 5–6 d before the egg is laid, whereas the white and shell reflect somewhat the body level of iodine in the hen at the time of laying. Thus, the yolk comes to equilibrium more slowly and, when the hen stops taking in iodine, the concentration in the white and shell falls off much faster than that in the yolk with a half-life of about 14 h.

8.4.2 Significance of dietary contamination levels

Strontium

The significance of the strontium content of the diet is assessed with respect to the working level calculated by the Medical Research Council; this is the average content in the diet over one year, indefinite exposure to which will cause the deposition of half the maximum permissible level (MPL; Section 6.1.1) in bone. Over the whole population, the working level is 130 pCi of Sr per g of Ca, and the MPL is 67 pCi of Sr per g of Ca in bone; for an individual the figures are 400 and 200, respectively. Since there is a long time interval between a nuclear detonation and deposition of the resultant strontium in bone, the situation should be reappraised, with a view to taking some remedial action, if the strontium in bone is found to be rising continuously and to have reached the cautionary level, $\frac{1}{2}$ the average MPL or 33 pCi/g Ca.

Following the peaks of weapons testing in 1959 and 1963-4, the corresponding peaks of strontium activity in the diet were 9 and 23 pCi/g Ca, or only a small proportion of the working level. Of more concern are certain 'special sites' in which heavy rainfall, soil conditions, and other factors combine to give relatively high contamination in milk, estimated at about 5–8 times the country-wide average. However, this is still below the working level. In view of the time between dietary contamination with strontium and the analysis of bones contaminated thereby, it is important to be able to assess bone levels from the amounts in the diet. It is difficult to obtain statistically reliable figures for strontium in the bones of young children (the critical population group) especially in special sites, since infant mortality is low, but the observed ratio (bone: local milk) in the first year of life appears to be about $\frac{1}{4}$. The maximum reported concentration of strontium in United Kingdom milk in 1964 was 217 pCi/g Ca (in Caernarvonshire), so that the maximum level in an infant continuously exposed to the most adverse local conditions would be about 54 pCi/g Ca, which would give a dose of 140 mrem/year. By comparison, the mean level in young children in any large area was not expected to be more than 10 pCi/g Ca, a dose of 30 mrem/year and, over the whole country, not more than 7 pCi/g Ca, or 20 mrem/year. Provided there are no more weapons tests after 1964, these figures would steadily fall. These estimates are intentionally pessimistic: in fact, the highest figure yet published for strontium in bone is 9·5 pCi/g Ca, for a 6 month old child who died in October 1959. Under the worst possible conditions, strontium-89 could give a dose of 25 mrem during the year following its deposition, compared with the national mean of 3 mrem.

Caesium

Caesium-137 is less hazardous than strontium-90: if equal activities are ingested, some tissues may receive 100 times more dose from strontium than that delivered to any part of the body by caesium. However, caesium, being a gamma emitter, can constitute an external hazard, and some Eskimos are reported to have suffered mild radiation sickness as a result of caesium deposition from Russian weapons tests in the Arctic. During conditions of high fall-out, when direct contamination of plants is the major source of both strontium and caesium, their ratio in the diet is fairly constant, and measurement of one can give an approximate guide to the quantities of the other. Radiation doses from caesium can be measured by whole-body counting of its gamma radiation. No working levels have been established for caesium in the diet, since concentrations of strontium would become critical first, and would give warning that caesium would then have to receive attention too. Since the biological half-life of caesium is much shorter than that of strontium, children do not receive much higher doses than adults. The doses to bone of young children from caesium during 1963 in mrem may be summarised and compared with those from strontium as follows:

Country-wide mean: Sr-90, 30; Sr-89, 3; Cs-137, 5; Total 38
Worst conditions: Sr-90, 140; Sr-89, 25; Cs-137, 40; Total 205

In all cases the doses are below the MRC limits, especially with caesium, which adds little to the 100 mrem/year of background; however, it must be remembered that the long-term effects of radiation probably do not have a threshold (Section 7.4.1).

Iodine

The maximum acceptable level of iodine-131 contamination is 200 pCi/l of milk, averaged over a year, or correspondingly higher concentrations over shorter times. During the 1961–2 weapons tests, the national mean total for the last quarter of the year was 17% of the maximum acceptable yearly amount; only in Wales did the concentration exceed the MPL (then 130 pCi/l), but the total was only 24% of the acceptable yearly figure. Tests stopped for most of 1962 but, during the last four months of this year, 14·6% of the MPL for 1 year was attained and, if such high levels had continued, consumption of fresh milk would have had to be curtailed.

8.4.3 Surveys of dietary contamination

Surveys of such contamination with radio-isotopes are undertaken in order to assess the radiation doses received by the population eating the diet and, though this is their prime purpose, they can also provide useful information on food chains. Surveys are best organised on a regular routine basis, on the principles that forewarned is forearmed and that prevention is better than cure: action may be taken as and when indicated by the survey results, to keep the situation under control and to prevent crises developing. Without regular surveying a crisis could arise (apparently) suddenly, probably leading to public panic, and there would have been no prior preparation for it, making remedial action much more difficult or impossible.

Surveys covering fall-out are both widespread, to find the average levels of dietary contamination, and localised, in areas which have exceptionally high contamination due to unusual local conditions. Such 'special sites' are usually small and may be easily missed in a broad survey, so that the primary problem in their study is to locate them. Routine local surveillance around reactors ensures that people living nearby are not exposed excessively; knowledge of local food habits is important. Special emergency surveys are mounted after accidents involving any sizeable release of activity; by their nature they cannot be planned in advance, but they must still be organised quickly, and an outline plan of campaign is advisable.

The planning of radiation surveys is like that of radiotracer experiments in that they must be well-designed and thought out beforehand if they are to yield the desired information in a useful and accurate form. Their object must be kept in view, for the analysis of foodstuffs for their content of radio-isotopes is not an end in itself. An additional aspect is that of public relations: depending on how well (or not) it is handled, a radiation survey may either reassure the public or encourage their fears.

The remedial measures to be undertaken against contamination depend on the situation, whether it is local or widespread, and transient or continuing. Their cost has to be balanced against the radiation hazards that are likely to arise if nothing is done. Two principles are of general application: first, most contamination, especially in the early stages of an emergency, will be in the crops from relatively unproductive areas; and secondly, a high standard of food hygiene, e.g. washing or removing the outer parts of vegetables, can reduce intake of isotopes appreciably. Local incidents are easily dealt with by importing food from uncontaminated areas, but this is not possible in continuing and widespread situations.

Incidents with iodine are, by nature, transient, and will cure themselves if the food is stored for a few months, either the contaminated fodder eaten by cattle or the milk from them, until the activity has declined sufficiently. Nevertheless, it was reported that, following the Windscale reactor incident in 1957, which released a sizeable quantity of iodine-131 into the surrounding area and contaminated all the milk in West Cumberland, many gallons of milk were thrown away as unfit for consumption, though had it been dried or condensed, or made into cheese, so that it could be stored until the iodine activity had decayed away, it might not have had to be wasted. Iodine can be removed from milk by ion exchange, but this has little point for the above reason; administration of stable iodine will reduce radio-iodine uptake by the thyroid, but must be carried out within a few hours.

If contamination with strontium is widespread, the most effective remedy is its removal from milk by ion exchange; the benefit would be in proportion to the amount of milk drunk, and would protect mostly infants and young children. Uptake of strontium from soil could be reduced only by drastic and expensive changes in the farming system. It is not possible to influence strontium in the diet by increasing calcium.

Caesium is less hazardous than iodine or strontium, and there is less need for remedial action against it. Milk can be decontaminated, but only modification of animal diets could remove it from meat.

REFERENCES

ABERG, B., and HUNGATE, F. P., (Eds), (1966). *Radioecological Concentration Processes*, Pergamon, Oxford, 1040 pp.

AHEARN, G. A., (1968). 'Phosphorus-32 Uptake by the Sea-cucumber *Holothuria atra*', *Biol. Bull. Woods Hole*, **134**, 367–381.

ANDERSON, J. J. B., and COMAR, C. L., (1968). 'Strontium Retention as a Function of Age in Dogs', *Radiat. Res*, **34**, 153–169.

ASHBY, W. C., BEGGS, J. V., KASTNER, J., OLTMAN, B. G., and MOSES, H., (1967). 'Ecological Dosimetry', *Science, N. Y.*, **155**, 1430–1432.

BONI, H., (1969). 'Caesium Retention in the Body Varies with Age and Sex', *Nature, Lond.*, **222**, 1188–1189.

CHARNELL, R. L., ZORICH, T. M., and HOLLY, D. E., (1969). 'Hydrological Redistribution of Radio-Nuclides Round a Nuclear-Excavated Sea-Level Canal', *BioScience*, **19**, 799–803.

COHEN, Y., and TADMOR, N. H., (1966). 'Root Tracing by Two-Layer Radio-Isotope Application', *Int. J. appl. Radiat. Isotopes*, **17**, 573–582.

EBEL, J. G., and COMAR, C. L., (1969). 'Use of Sr-Ca Ratios in Hair for Monitoring', *Hlth Phys.*, **16**, 205–208.

ECHO, J., and HAWKINS, D. B., (1966). 'Algal Influence on Radionuclides in Settling Ponds', *Nature, Lond.*, **209**, 1105–1107.

FUJITA, M., IWAMOTO, J., and KONDO, M., (1969). 'Variation of the Sr-Ca O.R. (Urine–Diet) in Man', *Hlth Phys.*, **16**, 441–447.

GARROW, J. S., (1965). 'Potassium-40 in Infants can be Related to Nourishment', *Lancet*, (Sept. 4), 455–458.

GOLDSMITH, W. A., BOLCH, W. E., and GAMBLE, J. F., (1969). 'Retention of Radionuclides by Panamanian Clays', *BioScience*, **19**, 623–625.

GUEGUENIAT, P., BOVARD, P., and ANCELLIN, J., (1969). 'Physiochemical Form of Ruthenium and Contamination of Marine Organisms', *C.r. hebd. Séanc. Acad. Sci., Paris*, (D), **268**, 976–979.

HARKNESS, D. D., and WALTON, A., (1969). 'Carbon-14 in the Biosphere and Humans', *Nature, Lond.*, **223**, 1216–1218.

ICRP, (1966). 'Review of Inhalation of Radio-isotopes', *Hlth Phys.*, **12**, 173.

IINUMA, T. A., ISHIHARA, T., YASHIRO, S., and NAGAI, T., (1969). 'Accumulation in Infants of Caesium-137 from Powdered Milk', *Nature, Lond.*, **222**, 478–480.

KALNINA, Z., and POLIKARPOV, G., (1969). 'Sr-90 Concentration Factors of Lake Plankton, Macrophytes and Substrates', *Science, N.Y.*, **164**, 1517–1518.

KLINE, J. R., and JORDAN, C. F., (1968). 'Tritium Movement in Soil of a Tropical Rain Forest', *Science, N.Y.*, **160**, 550–551.

LEVI, H., LINDEMAN, J., and FENGER, K., (1969). 'Strontium-90 in Human Bone', *Hlth Phys.*, **17**, 354–356.

MRC, (1966). *Assessment of Possible Radiation Risks to the Population from Environmental Contamination*, HMSO, London.

MARTIN, W. E., (1969). 'Bioenvironmental Studies of Radio-Safety Feasibility of Nuclear Excavation', *BioScience*, **19**, 135–137.

MORGAN, A., MORGAN, D. J., and BLACK, A., (1968). 'Study of Deposition, Translocation and Excretion of Radio-Iodine Inhaled as Iodine Vapour', *Hlth Phys.*, **15**, 313–322.

MORLEY, F., (1970). Private Communication.

PALMER, H. E., BEASLEY, T. M., and FOLSOM, T. R., (1966). 'Iron-55 in the Marine Environment and in People who Eat Ocean Fish', *Nature, Lond.*, **211**, 1253–1254.

POLIKARPOV, G. G., (1966). *Radioecology of Aquatic Organisms*, (Transl. Scripta Technica), North-Holland, Amsterdam.

PRESTON, A., and JEFFERIES, D. F., (1969). 'ICRP Critical Group Concept with Respect to Windscale Sea Discharge', *Hlth Phys.*, **16**, 33–46.

SPARROW, A. H., and PUGLIELLI, L., (1969). 'Simulated Fallout and the Growth and Yield of Vegetables', *Radiat. Bot.*, **9**, 77–92.

STERNBERG, J., LÉGAVÉ, J. M., and MARCIL, B., (1969). 'Radiocontamination of the Environment and Effects on the Mother and Foetus', *Int. J. appl. Radiat. Isotopes*, **20**, 81–96.

TURNER, G. A., and TAYLOR, D. M., (1968). 'Transport of Plutonium, Americium, and Californium in Blood of Rats', *Physics Med. Biol.*, **13**, 535–546.

VANDENHOEK, J., KIRCHMANN, R. J., COLARD, J., and SPRIETSMA, J. E., (1969). 'Effects of Pasture Feeding, Pasture Type, and Season on the Transfer of Strontium and Caesium from Grass to Milk', *Hlth Phys.*, **17**, 691–700.

WOLFE, D. A., (1967). 'Seasonal Variation of Cs-137 from Fallout in a Clam *Rangia cuneata*', *Nature, Lond.*, **215**, 1270–1271.

WOODWELL, G. M., (1969). 'Radioactivity and Fallout—the Model Pollution', *BioScience*, **19**, 884–887.

Chapter 9

SOME SUGGESTIONS FOR PRACTICAL WORK

9.1 INTRODUCTION TO PRACTICAL WORK

Even the most elementary study of isotopes benefits greatly if the theory is illustrated with some practical work. The equipment and facilities available for isotope work vary greatly in nature and extent in different places, and the level of courses and the time spent on them also vary, so that it is rarely possible to follow without modification a practical schedule intended for use elsewhere. The following suggestions for experiments to complement the theoretical part of this book are therefore given in outline only, without precise practical details, as ideas on which a course can be based according to the facilities and time available. It is good if all aspects of the subject can be covered, although it is not very practicable to illustrate Chapters 5, 7 and 8 in an introductory course using simple equipment. The experiments given are not by any means an exhaustive catalogue, but merely a selection of those of which the author has had some experience, and which will fit into a three-hour practical period. They are arranged as follows:

- 9.2. Introductory experiments: assembly of a G–M counting set, characteristics of a G–M tube, randomness and statistics of decay, resolving time;
- 9.3. Natural radioactivity: solvent extraction of uranium and half-life of extracted protoactinium, abundance of potassium-40;
- 9.4. Stable isotopes: water-exchange studies using deuterium oxide;
- 9.5. Physical properties of radiation: external absorption, bremsstrahlung, backscatter, and self-absorption of beta radiation, gamma scintillation spectrometry, residual phosphorescence, half-life;
- 9.6. Health physics: monitoring, decontamination, shielding;
- 9.7. Autoradiography: uptake of phosphorus-32 by roots, sym-

plastic movement from the leaf, movement in tissue which is largely parenchymatous, chromatograms;

9.8. Tracer experiments: incorporation of iron into mouse erythrocytes, uptake of phosphorus by mouse tissues, active transport across a membrane, ion balance in aquatic animals.

The chapter concludes with a table of physical data for some isotopes of biological use and interest, and two examples of correction for radioactive decay.

The main expense in starting up an isotope course is fairly certain to be the electronic 'hardware', though satisfactory gear can be obtained second-hand, sometimes for as little as a few pounds per hundredweight; a competent electronics technician to maintain it is also very useful. Several beta- and gamma-emitting isotopes can be had quite cheaply.

It is of great benefit for the class to learn something of planning experimental procedure (Chapter 4) by doing it themselves for some or all of the experiments to be carried out, discussing their plans with the instructor before starting the practical work. The laboratory manual or schedule should, of course, encourage thought in this direction rather than being just a recipe-book. To this end, pertinent questions are posed at various points in this chapter; the answers are given at the end. It is also useful to gain some practice in calculations and the solving of numerical problems. Note Section 4.5 on corrections to observed counts.

The primary concern in the radio-isotope laboratory is safety in working, since radiation and contamination hazards are so insidious, and it may be said that safe handling of radio-isotopes is much the most important thing to learn in any practical course of instruction in their use. The main points to put across are:

1. Use of protective clothing, laboratory coats, and gloves, and how to put them on and remove them (Section 6.6.3);
2. No mouth operations (Section 6.6.4);
3. Keeping activity in its proper place, avoiding spread of contamination, and tidy working (Sections 6.6.6 and 6.6.7);
4. Care in handling open sources: working on trays, wearing gloves, using double containers, etc. (Section 6.6.6);
5. Cleaning up and decontamination at the end of an experiment, and disposing of waste (Sections 6.6.8 and 6.6.9);
6. Wearing film badges as a check on radiation exposure (Section 6.6.10).

It is as well to avoid open sources of alphas, other than uranium and thorium; these, incidentally, should be treated with more respect than they sometimes get.

9.2 INTRODUCTORY EXPERIMENTS

A useful starting exercise is the assembly of a Geiger–Muller (G–M) counting set (Section 3.5) followed by determination of its characteristics, estimation of its resolving time, and the statistics of radioactive decay. In addition, demonstrations may be set out, showing different kinds of scalers, ratemeters, G–M tubes, scintillation counters etc., and containers used for transporting isotopes.

9.2.1 Assembly of a Geiger—Muller counting set

The gear is presented in pieces: G–M tube and holder, probe unit, scaler, high-voltage power supply, timer, and connecting cables. The student should identify the several items, connect them correctly, draw and fully label the block diagram, and learn and understand what all the controls do.

9.2.2 Characteristics of a Geiger—Muller tube

A suitable source is uranium oxide, sodium-22, or a medium or hard beta emitter. It is spread over a planchet, cemented with a very weak solution of perspex (or the transparent rapid-drying type of adhesive) in acetone or chloroform, and covered with cellulose self-adhesive tape or thin aluminium foil. It should give about 5000–10 000 counts/min. Check that the high-voltage controls are 'Off' before connecting the counting set to the mains—why? Switch the set to 'Mains' or 'Test' and allow to warm up for about $\frac{1}{2}$ h. Leave the high-voltage off. Place the source under the G–M tube and switch the set to 'Count'. Increase the high voltage slowly until the scaler starts counting, reset the scaler to zero, and record about 10 000 counts at this voltage. Increase the high voltage in 25 V steps and take a count at each, allowing the set to stabilise for 1 min in between voltage changes. Plot the count rates (counts/min) as they are obtained, and observe the appearance of the plateau; the results should appear as in Figure 3.3 (p. 56). If the scaler starts to race at any time, *AT ONCE* decrease the voltage, and in any case do not go more than 300 V above the starting threshold: it is emphasised that an over-voltage may ruin the tube. Background may be taken over the same voltage range, though it will be low in comparison with source counts.

G–M tubes can show a hysteresis effect, so it is preferable to increase the voltage steadily, rather than reduce it to repeat an observation or to fill in a gap. Should this be thought necessary, complete the observation of the plateau, and lower the voltage to zero; wait a few min, and then increase it again to the desired value. Remove the sample immediately after counting—why?

Reduce the high voltage when the tube is not being used and see that it is turned off at the end of the day's work. Do not connect or disconnect with the high voltage turned on, especially with transistorised gear, in which the circuitry is easily damaged. The working voltage is taken as being about 50 V above the starting voltage, or roughly in the middle of the plateau; it is noted for future use. Beginners sometimes fall into the error of thinking that the working voltage varies for different isotopes, but an understanding of how a G–M tube works at once shows that this is not so—why? The plateau should be at least 100 V long in a halogen-quenched tube, and 150–200 V in an organic tube. Its slope is usually expressed as percentage change in count rate per V change in applied potential, i.e.

$$\frac{\text{Slope}}{\text{Working voltage}} \times 100\%/\text{V}$$

and it should not exceed 0·1%/V.

The characteristics should always be determined when a new combination of tube and scaler is used, and should be checked routinely. A quick method is to count a source at 25 V and at 75 V above the starting voltage; the two counts should not differ by more than about 10%.

The only trouble with this experiment is that inexperienced beginners may inadvertently exceed the maximum voltage for the tube, even if they are warned not to, and if the tube is labelled appropriately; this will ruin certain types of tube and make them unfit for further use, so that the experiment may turn out to be rather an expensive exercise, especially if G–M tubes are in short supply.

9.2.3 Randomness and statistics of decay

Place the source in position below the G–M tube, and without moving it, (why?) take a large number of counts. With an activity of about 10 000 counts/min and taking $\frac{1}{2}$ min or 1 min counts, a statistically reasonable number of results can be accumulated within 1 h. Plotted as a histogram, they approximate to a normal distribution about a mean, and they are checked for correspondence with statistical theory which predicts that:

1. Standard deviation (SD) is approximately equal to the square root of the mean count.

$$\text{Exact SD} = \sqrt{\left(\frac{\text{Sum of squares of deviations from mean}}{\text{Number of observations}}\right)}$$

2. Mean deviation = 79·8% (4/5) of SD.
3. 31·7% (1/3) of deviations exceed 1 SD.
4. 4·6% (1/20) of deviations exceed 2 SD.

9.2.4 Resolving time

This can be estimated simply by counting two sources separately and together; two D-shaped pieces of filter paper soaked in isotope solution to give a count rate of about 10 000–20 000 counts/min, dried and covered with cellulose self-adhesive tape, are suitable.

Place one source under one side of the G–M tube and count it; lay the second alongside, without touching or disturbing the first, and take the 'together' count; then remove the first source, without disturbing the second, and count the latter alone—why? Calling n_1, n_2 and n_3 the observed count rates of the two sources alone and together respectively, the existence of resolving time is shown by the fact that n_1+n_2 is not equal to n_3. The corresponding true count rates that would be obtained if there were no dead time can be calculated from the correction formula. If a source gives n counts per second in a counter with dead time t seconds, and its true count rate is N counts per second, then the counter is insensitive for nt seconds per second, and it misses Nnt counts per second.

$$\text{Therefore, } N-n = Nnt$$

$$\text{Or, } N = \frac{n}{1-nt}$$

When using the formula, note that n and t must be in the same time units! The sum N_1+N_2 should now be more nearly equal to N_3, though it may not be exactly so (*see* 9.2.2 above). The corrections obtained using the formula may be checked against the correction table (Table 9.1) to see that it gives the same answer; a different resolving time will of course need a different table.

The resolving time may be calculated from n_1, n_2 and n_3; theoretically, background should be considered since it is subject to coincidence loss in the same way as source counts, but, as it is

Table 9.1. CORRECTION FOR A RESOLVING TIME OF **400** μs (ALL FIGURES ARE IN COUNTS/min

Observed Count	0	200	400	600	800
0	0	200	401	602	804
1 000	1 007	1 210	1 413	1 617	1 822
2 000	2 027	2 233	2 439	2 646	2 853
3 000	3 061	3 270	3 479	3 689	3 899
4 000	4 110	4 321	4 533	4 746	4 959
5 000	5 172	5 387	5 602	5 817	6 033
6 000	6 250	6 467	6 685	6 904	7 123
7 000	7 343	7 563	7 784	8 006	8 228
8 000	8 451	8 674	8 898	9 123	9 348
9 000	9 574	9 801	10 028	10 256	10 485
10 000	10 714	10 944	11 175	11 406	11 638
11 000	11 870	12 104	12 338	12 572	12 808
12 000	13 043	13 280	13 517	13 755	13 994
13 000	14 234	14 474	14 714	14 956	15 198
14 000	15 441	15 685	15 929	16 174	16 420
15 000	16 667	16 914	17 162	17 411	17 660
16 000	17 910	18 161	18 413	18 666	18 919
17 000	19 173	19 428	19 683	19 940	20 197
18 000	20 455	20 713	20 973	21 233	21 494
19 000	21 756	22 018	22 282	22 546	22 811
20 000	23 077	23 344	23 611	23 879	24 149
21 000	24 419	24 689	24 961	25 234	25 507
22 000	25 781	26 056	26 332	26 609	26 887
23 000	27 165	27 445	27 725	28 006	28 288
24 000	28 571	28 855	29 140	29 426	29 712
25 000	30 000	30 288	30 578	30 868	31 159
26 000	31 452	31 745	32 039	32 334	32 630
27 000	32 927	33 225	33 524	33 824	34 124
28 000	34 426	34 729	35 032	35 338	35 643
29 000	35 950	36 258	36 567	36 877	37 188
30 000	37 500				

For example, 7777 counts/min observed = 8006+177 = 8183 counts/min corrected. This is slightly too low, and the correction may be done thus: 8228–23 = 8205 counts/min. However, the error is only 22 in 8000, which is less than 1% and within the limits of experimental error and statistical variation of the count, and for most purposes it is not necessary to work out corrections any more exactly.

by comparison small, it may be ignored without much loss of accuracy. The true count N is made up of counts from the source and from background, so that $N_1 = S_1 + B$ and $N_2 = S_2 + B$.

Now, $S_1 + S_2$ must be equal to S_3, so that $N_3 = S_1 + S_2 + B$ and $N_1 + N_2 = S_1 + B + S_2 + B = N_3 + B$.

From the equation for N above, this is converted into terms of the observed counts, giving:

$$\frac{n_1}{1-n_1 t} + \frac{n_2}{1-n_2 t} = \frac{n_3}{1-n_3 t} + \frac{n_b}{1-n_b t}$$

where n_b is the observed background count.

This equation is then solved for t. It is simplified by omitting terms in t^2, which are taken as being very small. Cross-multiplication and cancelling out terms yields the result:

$$t = \frac{n_1 + n_2 - n_3 - n_b}{2(n_1 n_2 - n_3 n_b)} \quad \text{or (ignoring background)} \quad t = \frac{n_1 + n_2 - n_3}{2 n_1 n_2}$$

A binomial expansion yields a different result, which is for some reason more accurate:

$$t = \frac{n_1 + n_2 - n_3 - n_b}{n_3^2 + n_b^2 - n_1^2 - n_2^2} \quad \text{or} \quad \frac{n_1 + n_2 - n_3}{n_3^2 - n_1^2 - n_2^2}$$

The method sometimes seems to give an inaccurate result, though the reason why is not known.

Table 9.2. CONVERSION FOR COUNTS PER MINUTE TO COUNTS PER 100 SECONDS

Enter table with counts per minute; *read off* counts per 100 seconds in body of table.

Thousands	*Hundreds* 0	1	2	3	4	5	6	7	8	9
0		166	333	500	666	833	1 000	1 166	1 333	1 500
1	1 666	1 833	2 000	2 166	2 333	2 500	2 666	2 833	3 000	3 166
2	3 333	3 500	3 666	3 833	4 000	4 166	4 333	4 500	4 666	4 833
3	5 000	5 166	5 333	5 500	5 666	5 833	6 000	6 166	6 333	6 500
4	6 666	6 833	7 000	7 166	7 333	7 500	7 666	7 833	8 000	8 166
5	8 333	8 500	8 666	8 833	9 000	9 166	9 333	9 500	9 666	9 833
6	10 000	10 166	10 333	10 500	10 666	10 833	11 000	11 166	11 333	11 500
7	11 666	11 833	12 000	12 166	12 333	12 500	12 666	12 833	13 000	13 166
8	13 333	13 500	13 666	13 833	14 000	14 166	14 333	14 500	14 666	14 833
9	15 000	15 166	15 333	15 500	15 666	15 833	16 000	16 166	16 333	16 500
Tens: Add	0	17	33	50	67	83	100	117	133	150

Example: 7777 counts/min = 12 833 + 133 = 12 966 counts/100 s to nearest 10.

Table 9.3. CONVERSION FOR COUNTS PER 100 SECONDS TO COUNTS PER MINUTE

Enter table with counts per 100 seconds; *read off* counts per minute in body of table.

Hundreds	*Tens* 0	1	2	3	4	5	6	7	8	9
0		6	12	18	24	30	36	42	48	54
1	60	66	72	78	84	90	96	102	108	114
2	120	126	132	138	144	150	156	162	168	174
3	180	186	192	198	204	210	216	222	228	234
4	240	246	252	258	264	270	276	282	288	294
5	300	306	312	318	324	330	336	342	348	354
6	360	366	372	378	384	390	396	402	408	414
7	420	426	432	438	444	450	456	462	468	474
8	480	486	492	498	504	510	516	522	528	534
9	540	546	552	558	564	570	576	582	588	594

Example: 7777 counts/100s = 4620+48 = 4668 counts/min to nearest 10.

A correction table (9.1) for a resolving time of 400 μs is given, along with conversion tables (9.2 and 9.3) which make life easier when one-minute counts have to be converted to counts per second or vice versa.

9.3 NATURAL RADIOACTIVITY

Natural carbon-14 and tritium are usable only if highly sophisticated gear is available, in view of their low specific activity, but uranium, thorium, and potassium are quite easily handled.

9.3.1 Solvent extraction of uranium, and half-life of extracted protoactinium

This is in essence very like the experiments carried out by Crookes and others in the early studies of natural radioactivity. It shows the principles of radioactive decay: what are they?

A 25% w/v solution of a uranium compound, e.g. uranyl acetate in strong hydrochloric acid (3 volumes of concentrated HCl and 1 volume of water), is provided. Extract with an equal volume of an organic solvent such as amyl acetate or methyl isobutyl ketone; as soon as the liquids start to separate, run off the bottom layer and then pour the top layer into a liquid G–M tube. This operation must be done fast but safely. Take 20 s counts, starting every 30 s, thus leaving 10 s for recording the count and resetting the scaler. The top layer may alternatively, though considerably less efficiently, be counted in a beaker under an end-window G–M tube. Continue counting until there is no point in going on. When is this?

Plot log of count rate against time of starting the count, and from the resulting line (which should be straight, albeit with increasing scatter as the count rate diminishes) estimate the half-life of the extracted protoactinium-234m, and its decay constant. ($t_{1/2} = -0{\cdot}3010$/(gradient of graph), from Section 2.2.1; actual value 1·18 min; *see* Figure 2.4, p. 22.)

The solutions may be returned to their bottles after the experiment and re-used. Observe the precautions for use of a liquid G–M tube (Experiment 9.8.2).

9.3.2 Natural abundance of potassium-40

This experiment reveals the fact that ordinary potassium, and hence all living things, are radioactive, a somewhat startling discovery to the uninitiated!

Weigh out some potassium chloride, or other non-deliquescent potassium salt on a planchet, sufficient to cover it. Count under a shielded end-window G–M tube; get as large a count as possible. Take also a background count—why is correction for background essential in this experiment? From the difference in the two figures (i.e. the net source count) estimate the natural abundance of potassium-40, or its specific activity—what must be allowed for? Take the half-life of potassium-40 as $1{\cdot}3 \times 10^9$ years.

9.4 STABLE ISOTOPES: WATER-EXCHANGE STUDIES USING DEUTERIUM OXIDE

Deuterium as deuterated water is about the only suitable material to introduce the use of stable isotopes; the detection of others is very difficult. Deuterated water may be used to study water exchange in a small aquatic animal. The concentration of deuterium oxide in a sample is assessed by comparing its density with those of pure water (H_2O) and of pure 'heavy water' (D_2O) using a density gradient of liquid with which water does not mix.

Prepare a density gradient from benzene (sp. gr. 0·874) and chlorobenzene (sp. gr. 1·105) to cover the range 1·00–1·105, the densities of water and of deuterium oxide, respectively. Take a tall thin measuring cylinder, pipette 1 ml of chlorobenzene into the bottom, and then add equal volumes of mixtures of chlorobenzene and benzene, whose proportions of benzene are steadily and evenly increasing, i.e. (0·95 ml chlorobenzene+0·05 ml benzene) followed by (0·90 ml+0·10 ml) and so on—the flow from the pipette can be slowed by a piece of rubber tubing and a screw clip. The organic liquids should be pipetted with safety pipettes, not directly by mouth. What precaution must be observed during the preparation of the gradient?

Calibrate the gradient with a few known concentrations of deuterium oxide: inject a very small drop from a hypodermic at the top of the gradient, and note where it comes to rest.

Place some *Gammarus* (or other small aquatic animal) in a specimen tube, just large enough to contain them comfortably, and add just sufficient deuterated water to cover them. Take samples of the water at intervals and determine the deuterium oxide content, using the density gradient.

Some practice is necessary for success in this experiment and to determine the ranges of D_2O concentration etc. which work best.

9.5 PHYSICAL PROPERTIES OF RADIATION

9.5.1 External absorption of beta radiation

This experiment shows how beta radiation is absorbed by matter and how the maximum range depends on the maximum energy, so that two isotopes in admixture can be analysed.

Two pure beta-emitters are used, one soft, e.g. Ca-45 or S-35, and one hard, e.g. P-32. One planchet of the soft emitter is provided, and two of the hard emitter containing respectively sufficient to give about 10^4 and 10^6 counts per min, together with a mixture of the two isotopes.

Count the weaker hard source alone and with aluminium absorbers of increasing thickness placed above it, adjusting the counting time so as to obtain about 10 000 counts at each stage. When the count rate drops to about 1000 counts/min, change to the stronger source, repeat the counts with the last three filters used for the weaker source (why?), and continue counting with the thicker absorbers. Do not expose the G–M tube directly to the strong source when changing absorbers. Plot the results, corrected for resolving time, semi-logarithmically against absorber thickness, displaying them all on one graph.

Repeat the experiment with the soft-beta emitter until background count rate is reached. Is the count rate with no absorber the true zero absorber thickness count rate? If not, how is the latter found?

Having done the above, analyse the mixture by taking two counts: what are the conditions under which they are taken? Assuming that the mixture was made with the same solutions as the separate samples and that the soft source and the weaker hard source contain unit mass of isotope, calculate the ratio of the masses of the two isotopes in the mixture.

9.5.2 Bremsstrahlung from beta radiation

This follows on from the previous experiment. Shield the stronger hard beta source with lead, of slightly greater thickness than the maximum range of the radiation, and count it for a fairly long period. Compare with background. Since the beta particles must be completely stopped by the shield (by definition), any counts recorded are obviously due to something else.

9.5.3 Backscatter of beta radiation

A convenient source for this experiment is a ring of filter paper which is soaked in phosphorus-32 solution, dried, and stuck to a metal plate fitting round a G–M tube holder. The source and the window of the tube face in the same direction. How should the source be set up, and why? Place gradually increasing thicknesses of aluminium below the source, fairly close to it, and take a count with each. Plot a graph of count rate against thickness of aluminium and estimate the saturation backscattering thickness: about 0·2 of the maximum range. Replace the aluminium by thick layers (why thick?) of lead, iron, tin (element, not tinplate), caesium iodide, paraffin wax, copper, and anything else available, preferably elements. What should be kept constant and why. Plot count rate against atomic number.

9.5.4 Self-absorption of beta radiation

Dissolve 1 g of sodium sulphate in about 25 ml water, and label it with about 2 μCi of sulphate-^{35}S solution. Add sufficient warm saturated barium chloride to effect complete precipitation. Centrifuge, and remove the supernatant.

Add a small amount of ethanol or acetone to the precipitate, agitate it with a Pasteur pipette, and transfer the slurry in small quantities to a dozen tared planchets. Distribute it as evenly as possible over the surface of each planchet with a little ethanol, dry carefully under an infra-red lamp, and reweigh. Place a tissue under the lamp to prevent contamination of the stand or bench; turn off the lamp when it is not being used. The planchets should contain masses of $BaSO_4$ ranging up to about 50 mg/cm^2: on the standard 2-cm diameter planchet, this is up to about 200 mg in total. Measure the area and work out the equivalent thickness. Count each planchet under a thin end-window G–M tube, correcting for background and resolving time if necessary.

Plot: first, count rate against sample weight, and secondly, log of count rate per mg of $BaSO_4$ against sample weight; and extrapolate to zero thickness. Estimate infinite (saturation) thickness. Below this point, the relationship between increase in count and increase in sample thickness is adequately described by a logarithmic law, and the second graph should be a straight line; indeed, its straightness is a test of how uniformly the samples are spread over the planchets.

The experiment may be carried out with barium carbonate containing ^{14}C, and similar specific activities are suitable. It is possible to use one planchet only, adding portions of slurry successively, and drying, weighing, and counting after each addition.

9.5.5 Gamma scintillation spectrometry

To find the optimum settings of the counter for a given isotope, proceed as follows: With the source in position, the gate open, and the bias control at the minimum (5 V), slowly increase the EHT voltage until counts are recorded. Take 2 min counts at 50 V intervals until the count rate begins to increase rapidly. Do not exceed the maximum voltage for the photomultiplier. Repeat with the source removed. Then repeat the whole experiment at 20 V bias.

Plot net source (S) and background (B) counts against EHT voltage. Calculate S^2/B for each setting, and plot this against EHT voltage; the maximum point of the curve indicates the optimum value for the high voltage, i.e. that giving the maximum efficiency of counting of the source relative to background.

There will usually be little difference in the performance of the counter at the two bias voltages. The bias can therefore be set at a level to suit other factors, e.g. elimination of unwanted pulses.

To obtain the spectrum of a given isotope, proceed as follows: Set the gate width at 0·5 or 1 V and record the pulse spectrum by counting at every 0·5 or 1 V setting of the bias from the minimum (5 V) to the maximum (usually 30 or 50 V depending on the instrument). Do this for two or three isotopes with different gamma energies and plot count rate against discriminator setting for each one. A graph of the gamma ray energy against pulse height corresponding to it should give a straight line through the origin, if the spectrometer is working correctly.

9.5.6 Residual phosphorescence

Prepare a 'blank' counting vial, containing only scintillator solution and no source material. Leave it in the sun for a few minutes, and then count it at intervals over the next few hours; it should start off with a high count that slowly declines.

9.5.7 Half-life

One experiment has been described above (Pa-234m in Experiment 9.3.1). Half-lives of isotopes such as P-32 and I-131 may be determined by counting samples at every meeting of the class, and comparing the results with long-lived isotopes such as Na-22 or uranium, which are used to check that the counting efficiency of the gear does not alter from time to time.

9.6 HEALTH PHYSICS

9.6.1 Monitoring

This experiment, 'hunt the source', is a variation of the traditional party game.

Armed with a portable monitor, the student tracks down several sources that have been previously 'lost' around the laboratory or one area of it. They should be gamma emitters, fairly strong, so as to be reasonably easily detected at a distance.

A tray is provided, loaded with a variety of objects, some of which are active and some not, as it might be at the end of a tracer experiment, e.g. planchets, syringes, beakers, tissues, source bottles, etc. Monitor each object in turn and remove those which are active to a separate tray. Compare the performance of different monitors, if more than one is available.

9.6.2 Decontamination

Several different surfaces have been contaminated; endeavour to decontaminate them, trying to achieve some standardisation of method, by treatments gradually increasing in severity: cold water, hot water, soap, rubbing with a cloth, and scrubbing. Rubber gloves and an overall are essential. Note the count rate on a ratemeter after each treatment.

Some suitable surfaces are: unglazed tile, cardboard (stuck to wood), untreated softwood, varnished wood, painted wood, Formica, glazed tile, and linoleum. A suitable isotope is phosphorus-32; a solution is squirted on from a hypodermic syringe and needle. Its hard beta radiation is easy to detect, it has no gamma hazard, and it decays away reasonably rapidly if it should get into the wrong place. What does the experiment show?

9.6.3 Shielding

A source of metallic cobalt-60 of about 1 mCi is ideal; a G–M tube designed for gamma radiation is better than an end-window type. Set up the counter in line with the source; keep the latter in its lead pot and remove the lid only when taking readings—why? Investigate the relationship between the intensity of the radiation received from the source and the distance of the detector from it. Check the agreement with the formula $D = 6CE$ given in Section 6.3. If there is sufficient equipment, compare the dose rate, as recorded on a radiation monitor or ratemeter, with the count rate as recorded on the scaler, using the same type of detector tube if possible. How far is the comparison valid?

Having done this, set the counter and source at a constant distance apart (about 1 ft or 300 mm) and investigate the effect of different materials interposed between source and counter, such as wood, glass, tile or brick, and lead, using a constant thickness (why?), so far as possible.

Study the effect of altering the absorber thickness on the intensity of radiation, using water in a wide tank; increase the depth in $\frac{1}{2}$ in or 1 cm steps. Different thicknesses of lead may also be used. Plot the results semi-logarithmically, and draw in the theoretical line using the absorption coefficient (0·064 cm^{-1} for cobalt-60 radiation in water, 0·72 cm^{-1} in lead). How may any difference from the experimental line be accounted for (apart from experimental error)?

The external absorption experiment above shows that shielding beta radiation is much easier than shielding gamma.

9.7 AUTORADIOGRAPHY

The simplest introductory experiments are on the macro-scale using whole plants and chromatograms. From start to finish, they extend over some days, but the actual number of hours they occupy is not excessive.

An accurate and complete picture of the absorption, translocation, and distribution of a labelled tracer may be seen in plants after only the mildest of treatments. Radio-isotope is administered at one point or in one area, and its travels through the plant are followed autoradiographically. Leaves or stems may be treated by applying tracer solution to a small area or to several areas, either as droplets or as spray. High humidity, bright light and relatively high temperatures (80°F or 25°C) promote rapid uptake. Water cultures are best used for root treatment. The quantities of tracer used in each case are of the same order of magnitude. How long the experiment is carried on depends on the process being studied: initial absorption may take place within minutes, translocation requires a few hours, and redistribution may extend over some days.

Every compound has a characteristic translocation pattern in plants, which can be investigated by treatment for varying periods. This can only be successful, though, if movement of the tracer is stopped at the end of a treatment period. Thus quick killing followed by rapid drying between hot blotters may result in the artefact of xylem transport, whereas freeze-drying is slower but reveals a true picture. Press-drying may be used in class experiments for convenience and speed. Why should the plants be flattened?

9.7.1 Uptake of phosphorus-32 by roots

Runner bean (*Phaseolus*) or tomato seedlings, or buttercups, have proved satisfactory. Transfer to water culture 24 h beforehand.

Set up several plants in small containers (e.g. Coplin jars), supporting them with a slotted cork; do not kink or crush the stems, or the experiment is foredoomed to failure. Add a few drops of phosphorus-32 solution (high specific activity, as orthophosphate). Remove a specimen after treatment for (say) 10, 30, 60, and 120 min. Wash carefully in the sink (avoid radioactive drips when taking them there), blot and press, displaying roots and leaves as well as possible. Aluminium foil between specimens in the press prevents cross-contamination.

When all plants are reasonably dry and flat, mount on herbarium paper within an area 10 in ×12 in, (254 mm×305 mm), the size of the x-ray film; use the space as economically as possible, for the film is expensive. Stick the plants down with a trace (not more) of Gloy or gum, or with small pieces of cellulose self-adhesive tape. Set up the autoradiograph in the darkroom. Support the paper with thick card, plywood or hardboard. Cover the plants with a

thin sheet of cellophane (why?), place a sheet of x-ray film on top, and mark it (e.g. by a cut with scissors)—why?

Wrap the whole in black polythene or paper, place under *gentle* pressure (a sheet of foam rubber takes up any irregularities) and leave for a sufficient length of time (1–2 d).

Develop, wash and fix the film at a convenient time—only now may it be exposed to white light. Compare autoradiograph and plants, interpret and comment.

A single large leaf may be used instead of a whole plant, and calcium-45 instead of phosphorus-32.

9.7.2 Symplastic movement from the leaf

Carbon-14-labelled urea is applied to the leaf; urease converts it to carbon dioxide which enters the photosynthetic cycle and thence forms sugars. These then move symplastically from the leaf to all other parts of the plant.

Small-sized *Tradescantia* are suitable, put into water culture a few days beforehand. Make a ring of lanoline or vaseline about 1 mm in diameter over the midrib, or a cluster of rings over the major veins near the base of the leaf; these positions are well placed for phloem transport. Inject a droplet of labelled urea into each ring. After treatment times of (say) 1, 2, 6, and 24 h, wipe off the grease, cover the treatment spot with masking tape, and then press the plants and prepare autoradiographs as in the preceding experiment.

9.7.3 Movement in tissue which is largely parenchymatous

This can be compared with the vascular movement seen in the two previous experiments.

Cut cubes of about 1 in (25 mm) side from potatoes, and place on wet filter-paper. Make lanoline or vaseline rings in the centre top of each block; apply carbon-14-labelled urea to one set of blocks, and carbon-14-labelled 2,4-dichlorophenoxyacetic acid (2,4-D) to another. Cover and leave for periods of 1, 2, 4, and 8 d. Cut thin slices from all blocks, some horizontally and some vertically. Freeze-dry and prepare autoradiographs. It is important to have sterile apparatus and conditions in this experiment, otherwise the potato blocks go mouldy.

Other more sophisticated treatments are exposure of the plant to radioactive gas (e.g. $^{14}CO_2$) or spray, or injection by micro-needles or aphid mouthparts: after being allowed to feed on the plant, the aphid is beheaded, its mouthparts remaining in position.

9.7.4 Chromatograms

The starting material is a hydrolysate of the protein from *Chlorella vulgaris*, which are grown in $^{14}CO_2$ as their sole source of carbon: it is obtainable commercially. Prepare a chromatogram of the amino acids, either in (a) n-butanol: glacial acetic acid: water, 40 : 10 : 25 by volume, which will separate about 10 spots but has to run overnight (15–16 h); or in (b) methanol: water: pyridine, 20 : 5 : 1 by volume, which separates with speed (2–3 h) rather than efficiency. When the chromatogram is dry, prepare an autoradiograph. The class may start at this point and be supplied with the chromatogram and its autoradiograph. Align the two (having marked them when the autoradiograph was being prepared) mark the spots through on to the chromatogram with a pin or mounted needle, measure their length, and cut them out. Count them, and the spaces between them, with a thin end-window G–M counter or (better) a windowless gas-flow counter. Draw a histogram of the results, to full scale, and compare with the density of the spots on the autoradiograph.

The amino acids in the hydrolysate are shown in Table 9.4.

If the isotope being used is a very low-energy beta emitter, there may be considerable self-absorption of the radiation by the chromatogram paper itself, particularly with tritium. The technique of

Table 9.4. AMINO ACIDS IN THE CHROMATOGRAM HYDROLYSATE OF EXPERIMENT 9.7.2

Amino acid	*Approx. % conc. in hydrolysate*	*Approx.* R_f (×100) *Solvent* (a)	*Solvent* (b)
Histidine	0–2	11	28
Lysine	6	12	15 (Streak)
Arginine	5	15	16 (Streak)
Serine	2	22	42
Glycine	3	23	37
Aspartic acid	6	23	40
Threonine	2	26	50
Glutamic acid	7	28	48
Alanine	5	30	52
Proline	5	34	55
Tyrosine	4	45	50
Valine	4	51	65
Phenyl-alanine	4	60	60
Isoleucine	4	67	68
Leucine	9	70	70

liquid scintillation autoradiography may be useful: the chromatogram is clipped to the x-ray film and immersed in a liquid scintillator; the light photons, rather than the ionising radiation, then blacken the film. (Wilson, 1958.)

9.8 TRACER EXPERIMENTS

Biological tracer experiments are legion in number, complexity, and time taken. A few possibilities are given in this section.

9.8.1 Incorporation of iron into mouse erythrocytes (Thomson, 1966)

Mice have been injected 24 h previously with about 1 μCi of iron-59-labelled citrate, administered subcutaneously in the region at the back of the neck. Kill the animals, open the chest, and take a sample of blood on to a weighed planchet. Reweigh, spread the blood with a little kaolin-in-alcohol suspension, dry, and count under an end-window G–M tube. Assuming that blood volume in mice is 7·5% and that haematocrit is 45·4%, work out the percentage uptake of iron-59 into the red blood cells.

If a liquid G–M counter is available, take another sample of blood and treat it as in the next experiment below—what does this give a comparison of?

Other tissues may be sampled and digested also, and their uptake of iron determined.

9.8.2 Uptake of phosphorus by mouse tissues

A group of three mice have been injected one or a few days previously with about 1 μCi of phosphorus-32. Take samples of several tissues—blood, kidney, spleen, liver, muscle, bone—in tared beakers. Digest them with nitric acid: add 1 ml of concentrated acid to each, stand a few minutes, cover with a watchglass (why?), and heat on a hot-plate until the contents just boil, then remove and again allow to stand. Repeat, adding another 1 ml of acid if necessary, until the tissues have dissolved. Make up to a convenient volume (say 10 or 25 ml) and count aliquots in a liquid G–M tube. Count also a dilution of the original solution that was injected; it is diluted with 4% v/v nitric acid rather than with water (why?).

Calculate the percentage uptake of phosphorus into each tissue. Is it necessary to correct for decay of the phosphorus-32 between injection and counting?

Note the precautions for use of a liquid tube:

1. Pour liquids into it, since a touch from a pipette may break the very thin inner glass wall and ruin the tube.
2. When pouring into or out of the tube, wrap a tissue round it to catch drips. Do this also when pouring from a cylinder.
3. Rinse out the tube with water between samples, and take a background count immediately before putting each sample into it (why?).
4. Turn off the high voltage when taking the tube from its holder or putting it back.

9.8.3 Active transport across a membrane

Active transport is characterised by the ability of a tissue to establish and maintain a concentration gradient across it, and may be demonstrated with sacs of everted rat gut and calcium-45.

Prepare Krebs-bicarbonate physiological saline containing 0·5% w/v glucose, and label some with calcium-45 to give about 2000 counts/min from a 0·2 ml sample. Pipette about 10 ml into each of four conical flasks, gas with 95% oxygen-5% CO_2 (avoiding splashing!) and leave to equilibrate at 37°C.

Kill a rat with a blow on the head, open it, and cut the small intestine where it leaves the stomach. Flush the gut out with ice-cold saline, and remove it from the body. Thread it on to a thin glass rod, turn it inside out, and lay it out in a trough of ice-cold saline. Take 100 mm segments from each quarter of its length, and wash them out with a little labelled saline. Tie off one end, fill the sac so formed with more radioactive saline, and tie off the other end, leaving a long tail of cotton to make handling easier. Incubate one sac per flask at 37°C for an hour; then remove, blot dry, and take 0·2 ml samples from the sac and the flask. Spread the samples on a planchet with a little kaolin in alcohol suspension, dry, and count.

9.8.4 Ion balance in aquatic animals

Crustacea such as crabs or crayfish are suitable; either influx or efflux of sodium, potassium, or chloride ions may be studied.

For influx of an ion, place the animal in labelled water (sea-water or fresh water as appropriate). At intervals of (say) every $\frac{1}{2}$ h, remove it, wash it to remove activity on its surface, and take a small blood sample; then replace it in the active solution. Blood samples should be taken from a different spot each time. Alternatively, if a gamma emitter is being used, it is possible to count the whole animal, and blood sampling is unnecessary. The ionic concentration of the blood of a fresh-water animal is far higher than that of the water in which it lives, so that this experiment can also serve to demonstrate active transport.

For efflux of an ion, load the animal previously with isotope, either by injection or by immersion for several hours in a radioactive solution. Wash the animal well, and transfer it to clean water. Take blood samples at intervals as above, or count the whole animal.

If a fresh-water animal is being used, the ionic exchanges may also be followed by sampling the water, though this is not possible with marine animals since the changes in concentration in the water are very small.

Influx and efflux may be followed simultaneously by a double-isotope technique: load the animal with sodium-22 and immerse it in sodium-24, and count the samples in the appropriate channels of a gamma spectrometer.

Results with sodium-24 and potassium-42 should be corrected for decay. *See* p. 278 for worked examples.

9.9 TABLE OF PHYSICAL DATA OF ISOTOPES

Table 9.5 gives brief physical data for some isotopes of biological use and interest, and includes most of those mentioned in the preceding text. For simplicity, only the main emissions are listed and, for full details, a reference text should be consulted. The percentage figure refers to the percentage of disintegrations in which that particular radiation is emitted.

Table 9.5. PHYSICAL DATA OF ISOTOPES

Isotope	*Half-life* (a=*year*)	*Beta (negatron except where stated)*		*Gamma*	
		Energy (MeV)	%	*Energy* (MeV)	%
Barium-140	12·8 d	0·48	25	0·16	10
		0·59	10	0·54	26
		1·01	60	and others	
Caesium-137	30 a	0·51	95	0·662	86
		1·17	5	via Ba-137 m	
Calcium-45	165 d	0·254	100		
Carbon-14	5760 a	0·159	100		

Table 9.5. CONTINUED

Isotope	*Half-life* (a = *year*)	*Beta (negatron except where stated)* *Energy* (MeV)	%	*Gamma* *Energy* (MeV)	%
Chlorine-36	3×10^5 a	0·714	98·3		
		EC	1·7		
Cobalt-57	270 d	EC	100	0·122	88·8
				and others	
Cobalt-60	5·26 a	0·31	100	1·17	100
				1·33	100
Fluorine-18	1·8 h	Positron		0·51 from positron	
		0·649	97		
		EC	3		
Gold-98	2·7 d	0·96	98·8	0·412	95·8
		and others			
Hydrogen-3 (Tritium)	12·26 a	0·018	100		
Iodine-125	60 d	EC	100	0·035	7
Iodine-131	8·04 d	0·61	87·2	0·36	79
		and others			
Iridium-192	74 d	0·54	41·8	0·296	26
				0·308	29
		0·67	46·8	0·316	73
				0·468	47
				and others	
Iron-55	2·7 a	EC	100		
Iron-59	45 d	0·27	46	1·10	57
		0·46	53	1·29	43
		and others			
Magnesium-28	21·4 h	0·42	100	0·032	96
		2·87 from Al-28		0·40	30
				0·95	30
				1·35	70
				1·78 from Al-28	
Mercury-197	65 h	EC	100	0·077	19·3
Phosphorus-32	14·3 d	1·71	100		
Potassium-40	$1 \cdot 3 \times 10^9$ a	1·32	89		
		EC	11	1·46	11
Potassium-42	12·4 h	2·0	18	1·52	18
		3·6	82		
Rubidium-87	$4 \cdot 7 \times 10^{10}$ a	0·275	100		
Ruthenium-106	1 a	0·039	100		
		1·5–3·6 from Rh-106		0·51–2·9 from Rh-106	
Selenium-75	121 d	EC	100	0·14	54
				and others	
Sodium-22	2·6 a	Positron			
		0·54	90·5	0·51 from positron	
		EC	9·5	1·28	100

Table 9.5. CONTINUED

Isotope	*Half-life* (a=*year*)	*Beta (negatron except where stated)* *Energy* (MeV)	%	*Gamma* *Energy* (MeV)	%
Sodium-24	15 h	1·39	100	1·37	100
				2·75	100
Strontium-89	51 d	1·46	100		
Strontium-90	28 a	0·54	100		
		2·27 from Y-90			
Sulphur-35	87·2 d	0·167	100		
Tantalum-182	115 d	0·18	38	0·068	31
		0·36	20	1·12	33
		0·44	23	1·22	28
		and many others			
Technetium-99m	6 h			0·140	90·1
				and others	
Thulium-170	127 d	0·88	24	0·084	≈ 3
		0·97	76		
Tritium (*see* Hydrogen-3)					
Xenon-133	5·3 d	0·34	≈ 99	0·081	≈ 35·5
Yttrium-90 (Daughter of Sr-90)	64·2 h	2·27	100		

Natural decay series

Isotope	*Half-life* (a=*year*)	*Alpha* *Energy* (MeV)	%	*Beta* *Energy* (MeV)	%	*Gamma* *Energy* (MeV)	%
Actinium-227	22 a	4·94	1·2	OR 0·04	≈ 99		
Protactinium-234m	1·18 min			2·31	≈ 90	0·75 most	
						1·00 abundant	
						and others	
Radium-226	1620 a	4·589	5·7			0·188	≈ 4
		4·777	94·3				
Radon-222	3·825 d	5·48	≈100				
Thorium-232	$1{\cdot}41\times10^{10}$ a	3·948	24				
		4·007	76				
Uranium-235	7×10^{8} a	4·18–4·56				0·185	55
						and others	
Uranium-238	$4{\cdot}5\times10^{9}$ a	≈ 4·2	100				

9.10 EXAMPLES OF CORRECTION FOR RADIOACTIVE DECAY

In Section 2.2.2, it was calculated from the basic equations for the decay process that:

$$\log_{10}N_0 = \log_{10}N_t + \frac{0{\cdot}3010\,t}{t_{1/2}} \qquad (2.5)$$

N_0 is the desired corrected activity (count) at the reference time t_0; N_t is the observed activity (count) at time t, which is the time between the reference point and the time at which the sample was counted; $t_{1/2}$ is the half-life of the isotope.

When the equation is antilogged, the antilog of the term $0{\cdot}3010t/t_{1/2}$ is the factor by which the observed activity (or count) N_t must be multiplied to give the corrected activity (or count) N_0. This applies when (as is usual) t is later in time than t_0.

Example 1

Iodine-125; $t_{1/2}$ 60 d, t 26 d after t_0.

$$0{\cdot}3010 \times \frac{26}{60} = 0{\cdot}1304, \quad \text{antalog} = 1{\cdot}350$$

$$\therefore N_0 = N_t \times 1{\cdot}350$$

The reciprocal of 1·350 is the fraction of the original activity remaining, = 0·7406.

Sometimes it happens that N_t is determined earlier than the reference time. Here t is negative in sign, and the term $0{\cdot}3010t/t_{1/2}$ becomes negative too, so that its antilog is then to be divided into N_t to give N_0.

Example 2

Iodine-125; $t_{1/2}$ 60 d, t 16 d before t_0.

$$0{\cdot}3010\,t/t_{1/2} = 0{\cdot}3010 \times \frac{-16}{60} = -0{\cdot}08027, \text{ antilog} = 1{\cdot}203$$

$$\therefore N_0 = N_t \div 1{\cdot}203$$

or, $N_0 = N_t \times(\text{reciprocal of } 1{\cdot}203) = N_t \times 0{\cdot}8312$

'Fraction of original activity' is now 1·203.

If N_0 is taken as 100, then N_t will be the percentage activity left after time t, and $\log_{10} N_t = 2 - 0{\cdot}3010t/t_{1/2}$. Taking the figures from Example 1, $\log_{10} N_t = 2 - 0{\cdot}1304 = 1{\cdot}870$, and $N_t = 74{\cdot}0\%$ of the original activity remaining.

9.11 ANSWERS

9.2.2. To avoid inadvertently applying an excessively high voltage to the tube, which might cause serious damage.
To prolong the life of the tube.
See Section 3.2.3.

9.2.3. To keep counting geometry constant.

9.2.4. It is impossible to replace one source in exactly the same position if it is picked up and replaced later.

9.3.1. First, decay is exponential, and is characterised by a constant decay factor and half-life; and secondly, an element transmutes to another as it decays, and the daughter may be chemically separable from the parent.
When the count rate becomes constant.

9.3.2. Because the source count rate is not high.
The fact that the detector will not pick up all the disintegrations in the sample, i.e. is not 100% efficient.

9.4. Great care must be taken not to disturb the liquid below as each successive layer is added.

9.5.1. To relate the two sources to each other.
No, for three reasons, first, the cellulose tape on the planchet—check its thickness (about 10 mg/cm^2) by weighing a piece and measuring its area; secondly, the air space between source and counter: 1·3 mg/cm^2 per cm of air gap; and thirdly, the counter window, whose thickness is usually marked on it. Plot the results, therefore, with the apparent absorber zero thickness set in from the left-hand side, and extrapolate back to find the true zero absorber thickness count rate.
With no absorber, and with the thinnest absorber that absorbs all the radiation from the soft source.

9.5.3. Well above the bench, to reduce backscatter from the bench to a minimum.
Thick layers ensure saturation backscatter for each material.
The distance from the source to the top of the scatterer, to keep the geometry constant.

9.6.2. The importance of working on impervious surfaces, with open sources, as it is impossible to decontaminate a wooden bench if activity soaks into it.

9.6.3. To minimise the radiation dose received from the strong gamma source.
Only for detectors whose efficiencies are the same as those used in the experiment.

So that the relative shielding effects of different materials can be compared.

Difference is due to build-up obtained under broad-beam conditions (Section 2.5.2); the relation of build-up to thickness of absorber may be worked out.

9.7. To sharpen the autoradiographic image.

9.7.1. To avoid the specimen sticking to the film or reacting chemically with it.

For identification later of which specimen belongs to which film, and their alignment.

9.8.1. The relative counting efficiencies of the end-window and of the liquid tubes.

9.8.2. To avoid nitrous fumes and radioactive spray in the air of the laboratory.

So that its specific gravity and therefore self-absorption is about the same as that of the samples.

No, because the standard solution and the samples decay at exactly the same rate.

To allow for any residual contamination of the counter by previous samples.

REFERENCES

CRAFTS, A. S., and YAMAGUCHI, S., (1964). *Autoradiography of Plant Materials*, University of California Agricultural Experimental Station Extension Service, Manual 35.

DANCE, J. B., (1967). *Radio-Isotope Experiments for Schools and Colleges*, Pergamon, Oxford, 200 pp.

FRANCIS, G. E., MULLIGAN, W., and WORMALL, A., (1959). *Isotopic Tracers*, 2nd edn (o.p.), University of London Press, London, 524 pp.

THOMSON, R. A. E., (1966). 'Incorporation of Subcutaneously Administered Iron-59 by Mouse RBC', *Nature, Lond.*, **212**, 925.

WANG, C. H., and WILLIS, D. L., (1965). *Radiotracer Methodology in Biological Science*, Prentice-Hall, New York, 382 pp.

WILSON, A. T., (1958). 'Tritium and Paper Chromatography', *Nature, Lond.*, **182**, 524.

INDEX